THE MODERN
DETECTIVE

THE MODERN DETECTIVE

DETECTIVE

How Corporate Intelligence

Is Reshaping the World

TYLER MARONEY

RIVERHEAD BOOKS ▪ NEW YORK ▪ 2020

RIVERHEAD BOOKS
An imprint of Penguin Random House LLC
penguinrandomhouse.com

LIBRARY OF CONGRESS CATALOGING-IN-PUBLICATION DATA
Names: Maroney, Tyler, author.
Title: The modern detective : how corporate intelligence is reshaping the world /
Tyler Maroney.
Description: New York : Riverhead Books, 2020. | Includes index.
Identifiers: LCCN 2020018425 (print) | LCCN 2020018426 (ebook) |
ISBN 9781594632594 (hardcover) | ISBN 9780698147928 (ebook)
Subjects: LCSH: Business intelligence. | Private investigators.
Classification: LCC HD38.7 .M3657 2020 (print) | LCC HD38.7 (ebook) |
DDC 658.4/72—dc23
LC record available at https://lccn.loc.gov/2020018425
LC ebook record available at https://lccn.loc.gov/2020018426

Printed in the United States of America
1 3 5 7 9 10 8 6 4 2

BOOK DESIGN BY LUCIA BERNARD

FOR KAREN, HAZEL, AND VIOLET

CONTENTS

THE MODERN
DETECTIVE

PROLOGUE

On February 8, 2018, at 10:22 p.m., Donald J. Trump opened his Twitter app and name-checked a private detective. "Steele of fraudulent Dossier fame," the president wrote. "All tied into Crooked Hillary."

It was a remarkable occurrence: an American head of state publicly acknowledging the work of a private eye, in this case a former British spy named Christopher Steele.

Trump had reason to fear, and disparage, Steele. A veteran of MI6, the United Kingdom's foreign intelligence service, with expertise in Russia, Steele helps run a private intelligence firm in London called Orbis Business Intelligence.

In mid-2016, during the U.S. presidential campaign, Steele uncovered what he believed to be salacious and treasonous behavior by Trump, who was then one in a crowded field of Republican candidates.

The leads Steele and his contacts developed—about Trump's

TYLER MARONEY • 2

sexual proclivities (potential blackmail material) and what Steele suspected was evidence of Russian collaboration with Trump's campaign (potential impeachment material)—so alarmed him that he alerted the Federal Bureau of Investigation, with whom he had consulted on other projects.

Orbis's report—what became known as the Steele dossier— had been leaked to *BuzzFeed*, and although the FBI had been investigating Trump's alleged collusion with Russia since mid-2016, it had the effect of a violent explosion: political chaos and media obsession ensued.

Steele found himself in the central nervous system of global political discourse. Trump's detractors cite Steele's work when compiling lists of the president's failings. Trump's supporters label it slanderous liberal dirt.

(When, in March 2019, the report issued by Special Counsel Robert Mueller included no explicit evidence of collusion with Russia, Trump, although relieved, continued to blame Steele, arguing Mueller's "witch hunt" was spurred by the "phony dossier.")

Steele's first client on the project was an organization financed by Paul Singer, a wealthy hedge fund manager who commits abundant resources to funding Republican causes and candidates, but who was no fan of Trump's. (At the time, Singer supported Marco Rubio's candidacy.) Steele had been contracted through an American research firm called Fusion GPS.

This kind of investigation is often called "oppo" work, in which candidates for public office, constituents, or issue advocates fund research designed not just to understand their opponents' integrity and reputation but to undermine them to gain an advantage.

When, in late 2016, it became clear that Trump would be the Republican nominee, Singer ended the engagement, and Fusion GPS recruited another client to fund further work to chase the tantalizing leads: Hillary Clinton's campaign.

Steele and one of the founders of Fusion GPS, Glenn Simpson—a former investigative reporter for *The Wall Street Journal*—have joined the ranks of modern private detectives (whether revered or reviled) who might be considered well known: Jules Kroll, David Fechheimer, and Anthony Pellicano are some others.

In 1992, *The New York Times* called Kroll "the world's most famous gumshoe." In 2018, the *San Francisco Chronicle* reported that Fechheimer, who died in April 2019, was "the veritable dean of American private investigators." In 2018, *The Hollywood Reporter* called Pellicano "the private eye to the rich, famous and distressed." (In 2008, Pellicano was imprisoned for bribing police officers and wiretapping, among other misdeeds; he was released in March 2019.)

Steele's fame forced him underground for a spell, but it has been good for business. Orbis landed a slew of new clients, and it has been reported that the company received two thousand job applications soon after the firm's work was revealed.

Steele's notoriety brings into relief the crucial, if usually discreet, role private detectives play in modern society—smoothing the flow of global capital, resolving disputes among private citizens and corporations, and ensuring that our legal systems, government agencies, and financial institutions remain fair and transparent.

(Of course, not all clients have such benign motives for hiring private eyes.)

Why does anyone hire an investigator? To uncover wrongdoing. To right a real or perceived wrong. To punish or to exact revenge. To gain an advantage over a competitor. To satisfy a curiosity. To find a missing person or recover a stolen object. To feed paranoia. To benefit the public interest. In other words, to help satisfy basic human impulses.

When people or companies are harmed—by a crime, by intentional or accidental noncriminal conduct (what lawyers call a tort), by physical violence, or by reputational damage—there are scores of remedies, including the civil courts, law enforcement, government regulators, and media coverage.

Private investigators are often called upon to help find the information that will help attorneys litigate, prosecutors indict, regulators sanction, and journalists report.

We are everywhere. We work for large companies, government agencies, A-list movie stars, professional athletes, nonprofits, sovereign nations, media organizations, and business tycoons, among others.

We are engaged by frozen food processors, video game designers, diamond dealers, jet engine manufacturers, nut sorters, collateralized mortgage brokers, coffee bean shippers, book publishers, patent licensees, sports teams, railcar builders, miners, retail chain operators, university deans, plastic pressers, building contractors, labor organizers, and pretzel sellers.

Our fingerprints are on litigation battles, mergers and acquisitions, public relations campaigns, loan agreements, securities trading, labor campaigns, investments, executive hires, political campaigns, and front-page news stories.

Elon Musk, Alex Rodriguez, Hewlett-Packard, Uber, Credit Suisse Group, Jay-Z, Sean "Diddy" Combs, Drexel Burnham Lambert, Goldman Sachs, the government of Kuwait, Elizabeth Taylor, and Tom Cruise have all hired private eyes.

Even the most indigent among us benefit from the skills of investigators. Federal and state governments often not only provide defendants with lawyers but also subsidize investigators, among other experts, to support a legal defense.

In the United States, there are more than thirty-five thousand private investigators, with nearly four thousand in Florida alone, according to the U.S. Department of Labor. Many who join the trade come not with relevant university degrees (in criminal justice, for instance), but are refugees from other industries: law enforcement, journalism, accounting, the labor movement, government intelligence, software development, and academia, as well as nongovernmental organizations, think tanks, and law firms.

Corporate investigators—as distinguished from the lone-wolf gumshoe who lurks in the shadows with a telephoto lens—are often hired by wealthy investors, Fortune 500 companies, wealthy individuals, and global law firms to conduct due diligence in advance of transactions, provide support to litigators during high-stakes legal disputes, or uncover fraud within large corporations.

Each detective agency has its own DNA. Some shops run networks of contractors around the globe. Others have a narrow focus—in a region (Southeast Asia, say) or a technology (data scraping for disinformation online, for instance).

Still others are all things to all clients: providers of background checks, executive protection, forensic accounting, criminal defense, surveillance. Some market a single skill, such as political opposition research or tracking down missing assets.

My firm, QRI, specializes in large, complex investigations that seek to benefit the public interest: reforming the criminal justice system, holding corrupt government officials accountable, exposing financial crimes, fighting causes of the climate crisis.

Each American state regulates the detective industry in its own way. Where I live, the New York State Division of Licensing Services permits private investigators to "obtain information regarding the identity, habits, conduct, movements, whereabouts, affiliations, associations, transactions, reputation, or character of any person or group of persons."

We can investigate "crime or wrongs done or threatened against the government of the United States of America or any state or territory of the United States of America," and we can assess the "credibility of witnesses or other persons" and identify "the whereabouts of missing persons."

We can gather information about "the location or recovery of lost or stolen property" and "the causes and origin of, or responsibility for, fires, or libels, or losses, or accidents, or damage or injuries to real or personal property or the affiliation, connection or relation of any person, firm or corporation with any union, organization, society or association, or with any official, member or representative thereof."

We can collect "evidence to be used before any authorized investigating committee, board of award, board of arbitration, or in the trial of civil or criminal cases."

We can obtain information "with reference to any person or persons seeking employment in the place of any person or persons who have quit work by reason of any strike" or "with reference to the conduct, honesty, efficiency, loyalty or activities of employees, agents, contractors, and sub-contractors."

As a corporate private detective, I have access to a broad spectrum of business and legal proceedings, transactions, tactics, behavior, strategies, and personalities from which I have absorbed lessons that are not taught in MBA programs or law schools and cannot be learned from reading *The Wall Street Journal* or the *Financial Times*.

These lessons were gleaned from my work as a PI and from investigators I met around the world while reporting this book, an experience that came with a challenge (that I also encounter in my day job): the misperception that private eyes are amoral rule flouters who deploy dark arts—hacking, impersonating, bribing. Popular culture is replete with such narratives, both real and fictional.

My colleagues and I work hard to maintain integrity, and I would argue that private investigators benefit the public good in sharp contrast to our reputations. We are traditionally hired by criminal defense lawyers to counter the unparalleled resources of prosecutors; we prove the existence of financial frauds for companies in cases where law enforcement is uninterested; we expose civil rights violations; we recover money squirreled away offshore by despots and dentists alike.

But the media is fascinated with the archetype of the rogue detective. I have a Google alert for terms like "private investigator" and "private eye," and most of the links that land in my in-box are to stories about investigators running afoul of the law: "Former Private Eye Pleads Guilty to One Count of Wire Fraud"; "Who's Investigating Fake Chinese Goods? Fake Investigators"; "Private Eye Cops to Bribery Scheme with NYPD Officer."

In 2017, private eyes were, once again, dragged from the shadows, exposed as miscreants, equivocators, and worse. An Israeli firm called Black Cube, among other companies, was hired by the film producer Harvey Weinstein through his lawyer David Boies to gather compromising information on women Weinstein had allegedly victimized (such as Rose McGowan) and journalists whose articles Weinstein sought to quash (such as Jodi Kantor).

Black Cube, it was reported, employed classic techniques of the clandestine services from which some of its employees had come: fake identities, fictional companies as cover, pretexting, lying, deceiving.

If this is true, these activities might have been criminal—a disservice to Black Cube's clients, to the industry, and to society. (In 2016, two Black Cube employees were convicted in

Romania for intimidating a prosecutor and hacking into her emails.)

In 2016, it was revealed that Uber had hired retired spies from a firm called Ergo (whose slogan is "Intelligence for an opaque world") to investigate someone who had sued the company. It is common in corporate America to hire investigators to assess the credibility of legal adversaries. What is uncommon were the firm's tactics.

Among critiques of Ergo's work by the New York district court judge Jed S. Rakoff, who oversaw the case in federal court, was that its investigators were not licensed (a violation that, at least in New York, where Ergo operated, can be a misdemeanor), and they lied about why they contacted witnesses. Rakoff was not circumspect in his opinion of this behavior, labeling it "blatantly fraudulent and arguably criminal."

The tool kit available to private investigators is considerably less potent than the one available to spies and cops and prosecutors. We cannot flip witnesses, blackmail agents, develop confidential informants, bug phones, offer protection, send subpoenas, or bribe sources. This means we must, almost by definition, be more creative.

Private investigators are occasionally allowed to "dissemble"— not tell the whole truth. When, for example, we are working to uncover violations of civil or intellectual property rights, the dissemblance is authorized by law, the evidence sought "is not reasonably and readily" available through other lawful means, or the dissemblance does not violate the rights of third parties.

Many investigators disregard such restrictions. Others operate in countries where such limits do not apply. Some exploit shady reputations to attract clients with shallow morals and deep pockets.

In 2005, after a decade in journalism, I joined Kroll Associates, the seminal corporate investigative firm, and became a private detective. The prospect of my new profession was thrilling, but I was apprehensive. I had not studied law or finance or accounting or criminal justice. I was not a retired cop, intelligence operative, prosecutor, or litigator. I had no experience mining data or employing digital forensic tools. I didn't understand how the SWIFT financial messaging network worked or how to get behind a website whose registrant was hidden behind a proxy.

Although I have been at it for more than fifteen years now, I am still in awe of the depth and volume of issues my industry, broadly defined, tracks. A trade publication called *Intelligence Online*, which chronicles gossip, trends, and news on global intelligence agencies, security firms, and corporate investigative companies, reminds me of this.

One recent issue covered freelance Brazilian armaments advisers, "gas diplomacy," the expansion of private security squads in the Emirates, "e-policing" in Poland, innovations in Israeli missile designs, Scandinavian "offensive" Trojan horse computer hacking firms, telephone communications interception tools, an examination of where fallen dictators hide assets offshore, diplomatic cable leaks, identity-anonymizing software, "fleets" of antipiracy developers, cyber defense tactics, and master code breaking.

We live in an era marked by a kaleidoscope of disarming trends: technological innovation, rising populism, new brands of autocracy, "fake news," deregulation, demographic division, and eroding public discourse.

Governments employ, and are victimized by, cyber espionage. Vast tranches of private consumer data are stolen by shadowy hackers and global technology companies alike. Social media platforms are poisoned by disinformation campaigns. Artificial intelligence and facial recognition technologies are growing in scale and accuracy. A free press is under attack. Amateur drones buzz above us.

If these developments feel chaotic and unnavigable, my experience as a private detective, as a professional fact finder, as an analyst of hidden information, has given me the optimistic perspective that once in a while our work—exposing corruption, providing intelligence to corporate clients, serving the indigent, reforming bureaucratic rot, uncovering fraud, and solving riddles—provides a measure of relief.

Each chapter in *The Modern Detective* brings the reader behind the scenes of a different investigation, reveals some tradecraft (the elusive *how* of our field), profiles some eccentric characters, and offers impressions on how the business world really operates—all from the perspective of practitioners.

This book is an attempt to share a few of the lessons I have learned: for instance, a lawsuit can be used not just to right a perceived wrong in the theater of the courts but to air dirty laundry about an opponent, to tell one side of a story, or to force the disclosure of hidden information; not all news stories have roots in the minds of reporters, and many journalists fail to independently corroborate what is fed to them by sources; our criminal justice system is broken and biased.

This book is also an attempt to express my enthusiasm for this industry and to argue that as players in global commerce, political battles, and international litigation, we have become indispensable to those systems.

(Some names and dates have been changed, and some narratives and characters are composites.)

1.

DOORSTEPPING

In the fall of 2016, I drove to Paterson, New Jersey, to meet with a congenital criminal. Bill Antoni's rap sheet ran twelve pages: drug possession, assault, burglary, attempted manslaughter. He was also a former confidential informant for the Paterson Police Department who had recently been released from prison.

I did not call Antoni or try to make an appointment. Instead, I went to his apartment building unannounced—what private eyes call doorstepping. It is easier to hang up the phone on someone than it is to slam the door in her face.

I also wanted Antoni to see me so he knew who was asking him to recant sworn statements he'd made to police detectives twenty years ago. I was hoping he'd sense my credibility, or at least give me the opportunity to prove it.

His apartment building's intercom did not seem to work, so I waited for a neighbor to come out of the building to let me in.

The tinny door clanged behind me as I pushed through a scrum of children in the foyer, their squeals echoing off the vaulted marble ceiling.

As I neared the top of the second flight of stairs, Antoni appeared, blocking the staircase. I recognized him from his mug shots.

"Mr. Antoni," I said. "My name is Tyler Maroney. I'm a private detective. I'm here to ask for your help on a case."

He was listening. A good sign.

"My client is a man who spent more than ten years in prison for a crime he did not commit. He was a victim of police misconduct, and you may have information that can help. He filed a lawsuit against the city and the cops that he hopes will lead to compensation for the injustices he suffered."

I had rehearsed this approach—an attempt to deputize instead of antagonize him. I was gambling that Antoni sought penitence. He was from a broken home, a former drug addict, in his late forties, on parole. Perhaps he was exhausted by a life spent estranged from his family and in custody.

"Who is your client?" he asked. His question suggested a willingness to cooperate.

I told him (I cannot disclose his name) and then explained to Antoni how I had come across his name in the transcript of a hearing from 1996 when Passaic County, New Jersey, prosecutors were considering calling him as a trial witness.

"How did you know I lived here?" he asked, more out of curiosity than defensiveness.

I had anticipated this question, too. "Through database research," I said.

My first exposure to the tools of powerful data mining that are available to private investigators was through a product called Accurint, whose predecessor, Seisint, for "seismic intelligence," was developed by Hank Asher, a former cocaine smuggler, housepainter, gin drinker, and computer programmer who died in 2013 and whose technology is credited with helping law enforcement agencies find terrorists, missing children, pedophiles, and murderers—and with making him rich.

In 2004, *Vanity Fair* wrote that Asher's tool helped investigators track down John Allen Muhammad and Lee Boyd Malvo, two serial-killer snipers who, over ten months in 2002, shot and killed seventeen people and injured ten more, becoming known as the Beltway snipers. Using Accurint, investigators drew up a list of all of Muhammad's "relatives, associates, every place he ever lived," Asher told the magazine. "The next morning they were in Tacoma, Washington, cutting the stump down that had the projectiles"—bullets—"in it that matched the projectiles from the killings." The snipers were nabbed by a SWAT team the next day.

I learned about Accurint in 2005, when I joined Kroll. That year, I turned thirty-three, was recently married, and had made a career change. During the previous decade, I had been a journalist: on staff at a national magazine, working as a documentary film producer, and contributing articles to various magazines and newspapers as a freelance reporter. I wrote a cover story for *ARTnews Magazine* about a frenzy for Andy Warhol and an article for *The Atlantic* about a diplomatic row between England and Spain over a nuclear submarine that had a reactor meltdown in the Mediterranean.

As a reporter for *Fortune* magazine, I covered technology start-ups, economics, and investing. I interviewed Jesse Jackson about discrimination on Wall Street, Dr. Dre about Napster, Bill Bradley about community service. I spent a weekend at the Bellagio covering the Disneyfication of Las Vegas. I wrote about a travel website run from Havana by an exiled former CIA officer whose exposure of American intelligence agents led to the agents' murders.

I had some small scoops. I was among the first journalists to tell the world about a travel website called Orbitz. I broke the news that the married founders of Nerve.com, a website from the dot-com era that peddled "literary smut," were separating.

I lived in Spain on a Fulbright fellowship for professional journalists. My byline appeared in *The New York Times* and *Sports Illustrated*. I helped produce two documentary films for *Frontline*, one of which won an Alfred I. duPont–Columbia University award, the broadcast and digital news equivalent of a Pulitzer.

But for all my success, I sometimes felt insecure, especially when covering the business world, and as a generalist I never quite knew where to find my next story. I had not studied finance and did not know a proxy statement from a quarterly report. I used acronyms that baffled me: OTC, EBITDA, P/E. I could not have told you the difference between a limited liability company and an S corporation.

During these years, however, I became enthralled with investigative journalism, which had a fundamental magnetism for me: uncovering hidden information. So I wrote a few investigative pieces. For instance, for *The New York Sun*, a now-defunct newspaper, I exposed fraud and discontent at the Greenmarkets, New York City's farmers markets.

But I had not inspired an editorial rant, a congressional inquiry, a regulatory probe, or a criminal investigation. Meanwhile, the media industry was reinventing itself. Traditional brands collapsed, print went digital, and advertising and subscription revenues dried up. My foray into muckraking led me to consider private investigative work. The industry, I hoped, would pay better than the media and offer opportunities for growth.

What I was looking for was a way for my work to matter to people—and to myself—and uncovering wrongdoing offered that promise. A chance encounter led me to apply to Kroll, where I was offered a job as a senior analyst.

To my ear it sounded spooky. During my first week, my insecurities grew. My colleagues came from the CIA, various district attorneys' offices, the FBI, the faculty of Ivy League business schools, the diplomatic corps, and Big Four accounting firms.

When I emailed an executive of the firm to ask where his office was located so I could introduce myself, he refused to tell me. "You're an investigator now," he said. "Find me."

Jules Kroll is legendary—the protagonist in the creation myth of the modern private investigator who's been featured

TYLER MARONEY ▪ 18

on *60 Minutes* and profiled in *The New Yorker*—but there are conspiracy theorists who argue that the company's detectives were agents of the New World Order and the Illuminati. One blogger spreading disinformation accused Kroll of "creating a 'massive virtual computer infrastructure' linking the Governments and intelligence communities of nearly every civilized nation on earth with one solitary purpose, since 1966. Covering the sins and Planet-Wide Machinations of Rockefeller Inc."

There are less absurd and spurious ways to assess my alma mater. In the 1980s, the media dubbed it "Wall Street's private eye," and the firm is famed for tracking down the assets of the former Haitian president Jean-Claude Duvalier, Ferdinand and Imelda Marcos, and Saddam Hussein.

Over the generations, thousands of Krollies, as some people in the industry call us, have left and formed our own companies.

A few days before I visited Antoni in Paterson, I had logged in to a database called TLO (a successor, of sorts, to Accurint). These databases—IRB, Clear, and Tracers are a few others—are owned by data giants like Reed Elsevier, Bloomberg, and Thomson Reuters.

As a PI, if I had to choose one digital resource to pay for, it would be this one. It is invaluable and relatively cheap—about $10 for a report on one person. And it led me to Antoni.

Documents from the civil rights law firm that hired me on

this case provided me with the background on Antoni, which is what made him such an attractive prospective witness.

While in prison in the 1990s, he had apparently overheard our client discussing the murder for which he was later convicted. But Antoni disappeared from the record after that, which is why the law firm, which sues government employees and agencies on behalf of victims whose civil rights have been violated, asked me to interview him.

Antoni glanced around the hallway and motioned for me to come into his apartment.

A black leather couch was shrink-wrapped in plastic. The moldings around the doorways and windows were blunted by decades of paint. A security camera was affixed conspicuously on the wall of the kitchen facing the only window with a fire escape. A holstered Glock was on the living room table. An FBI badge hung from a nail above the couch. A crucifix hung at a compass angle above the front door. Taped to the wall in an arc that swept across the living room were black-and-white printouts of photographs: the grassy knoll in Dallas, the Oval Office, Jackie O., Hyannis Port.

Antoni—wearing black slippers, black sweatpants, a white sleeveless T-shirt that exposed inked arms, a gold chain, and slicked-back hair—noticed me notice his shrine.

"Kennedy was the last great American," he said.

I knew he would talk.

As Antoni and I settled into his living room, I noticed that his television was on with the volume turned low. Pornography.

Because he was answering my questions, I did not point out how surreal this was. He soon realized, however, and, hardly embarrassed, reached over to hit the power button. The video went dark, but the audio remained. We spoke for an hour to a moaning Muzak.

Here is what he told me. While Antoni was in prison, an undercover officer approached him one afternoon and showed him a single photograph of one of their suspects. Antoni was told that he was to identify this man at trial in return for leniency on the drug charges he faced. Detectives from the Paterson Police Department abandoned their first plan for Antoni—to have him say he witnessed the shooting—because he had been in prison that day.

This was not unfamiliar territory for Antoni. He had a kind of side career as a professional informant, both offering information and agreeing to fabricate it. Antoni agreed to participate in the corruption—to testify that he had overheard the defendant confessing to murder while in central booking.

A few months later, when he was out on bail, Antoni and the undercover cop sat in a police cruiser in the parking lot of a Dunkin' Donuts while Antoni rehearsed a handwritten script the detective showed him. At the end of the meeting, the detective gave Antoni $500 in cash.

Antoni's credibility and memory were so marred by drug abuse and mental illness, however, that prosecutors declined to call him. But I now knew how the trap had been set, and I set out to find other witnesses who could corroborate what Antoni told me.

Antoni was remorseful. He agreed to provide an affidavit or testify to what he told me to help the man he helped convict.

"I just wanted to get out of jail," he said. "That kid was innocent and needed help," he continued. "But I fucked him." (Antoni's willingness to speak with me is rare: as a private detective, I can never compel a witness to speak; only law enforcement officers have that power.)

About a year later, the City of Paterson settled the lawsuit brought by the man Antoni helped imprison for about $10 million. The law firm that funded the investigation, for which we were paid about $75,000, litigated the case for years, and its fee was approximately one-third of the $10 million.

Antoni was also looking for some kind of redemption, I suspect. However, our conversations helped a man who had spent years in prison for a crime he did not commit earn some measure of justice: if not an apology by the government agencies that conspired to steal his freedom, then at least money and self-respect.

The business model of the law firm that filed this case is, in part, to earn revenue from settlements or jury awards with government agencies. But there is a greater good in this kind of legal strategy.

This kind of litigation, supported by investigators like me, helps hold our elected and appointed officials accountable for overseeing a system that is rarely concerned with reforming itself.

This case also illustrates the reach of "big data." The databases I used to find him are obscure but available, in various iterations, not just to private investigators but also to insurance companies, law enforcement agencies, law firms, intelligence agencies, government contractors, health-care services companies, media organizations, and banks, among other customers.

A typical report includes your first, middle, and last names, as well as other names you've used on an official document like a utility bill or a credit card application or a property deed; name variations; your date of birth; your Social Security number; the state your Social Security number was issued in; the approximate date your Social Security number was issued; the names of other people who have been associated with your Social Security number; home phone numbers; mobile phone numbers; phone numbers you have used previously, landline and wireless carriers; your address and every address linked to you during the past generation; the counties in which those addresses are located; the approximate dates you have been linked to those addresses; the names of the owners of every address you have ever been linked to; your driver's license number and the state in which it was issued; the cars you have registered in your name, along with the model, make, and vehicle identification numbers of those cars and the dates those cars were registered; your current and former employers and other companies linked to you; professional credentials (such as an accounting or law or pilot's license); planes you have owned; boats you have owned; your voter registration information, including the state in which you are registered; your party affiliation and the date of your most recent vote; any hunting and fishing permits you have; tax liens placed on you; civil judgments against you; Uniform Commercial Code financing statements naming you; the names, states, and addresses of companies you have set up; the names of other people who helped you set up those companies; the names of your close relatives, the names of your neighbors, and the names of your neighbors from every place you have lived in the past generation; the block, lot, and parcel number of every one of those addresses; the assessed value and market value of every one of those properties; the census information on any addresses linked to you and your neighbors and neighborhoods; the dates the buildings you lived in were built; the most recent sale price of every one of those properties; how much you paid for any piece of real estate you have bought; the amount of any mortgage you took out to make those purchases; the names of the people or companies that sold you those properties; the names of the people or companies who bought those properties from you; certain criminal records; and your bankruptcies.

Not all of this information is found in every report, of course. And it can be inaccurate. All of it is bought, scraped, and sorted from thousands of sources, both public and private: court archives, property recorders' offices, utility companies, consumer credit agencies, telephone companies, social and business networking sites, secretary of state offices, professional licensing bodies.

The information in the reports is governed by federal and state laws such as the Gramm-Leach-Bliley Act, the Fair Credit

Reporting Act, and the Driver's Privacy Protection Act, and users must certify that they have a legitimate purpose to access it, such as supporting a legal proceeding, resolving a customer dispute, or protecting against fraud,

It was while protecting against fraud that a year earlier I found myself in midtown Manhattan with a private investigator named Louis.

2.

HASHING

After midnight on a muggy summer Friday, a private eye named Louis and I walked down Lexington Avenue. I was empty-handed, but Louis carried a crush-proof, watertight Pelican briefcase molded from "Ultra High Impact" structural copolymer resin.

The case was designed to transport—in open-cell foam core—rifles, laptops, shotguns, high-caliber ammunition, listening devices, and cameras, among other matériel.

Pelican cases are "NATO codified and tested" to "Military Standard" and are corrosion-proof and dustproof. They have survived lion attacks in Zimbabwe, tsunamis, C-4 blasts, and helicopter gunship crashes. (Pelican guarantees state that its cases do not cover "shark bite, bear attack or children under 5.")

Louis was not wearing fatigues. Instead, he wore a maroon Gap button-down shirt, pleated khakis, and black Reeboks. As he approached the lobby of an office tower near Grand Central

Terminal that held the headquarters of an investment firm called Phillip Solomon Partners, or PSP, he realized that for all the vital paraphernalia in his airtight, chemical-resistant Pelican case, he had forgotten to pack White Chocolate Macadamia Nut Clif Bars.

Miffed, Louis walked past the PSP building to find an open deli. It was going to be a long night, and he needed his snacks.

The normally chaotic streets of Manhattan were calm. The primary colors of corporate brands—Chase, GNC, Starbucks—gleamed from the closed retail shops at the base of the glass and steel canyon that is Lexington Avenue. Yellow cabs, Ubers, and Seamless deliverymen on electric mopeds enjoyed the unclogged four-lane street. The blinking streetlights cascaded downtown into the darkness.

In the Art Deco lobby of the office building that housed PSP, Louis (not his real name) and I were met by Sally, a PSP secretary who vaped. Sally was surly about having to stay late to ensure that all ninety employees had left before admitting me and Louis. She had been instructed by both her boss and PSP's general counsel to give us anything we wanted, including access to a corner suite on the nineteenth floor, which, during daylight hours, was occupied by Duane Park, a PSP managing director.

Park's office was unremarkable. His desk was cheap blond pine, and the windows offered a view of another skyscraper, not a dramatic cityscape or the horizon line. The ceiling lights were sterile fluorescents and the carpet Ikea-grade with a generic graphic print.

But the dozens of photographs arranged on his desk and covering the walls revealed an opulent life. There he was on his sixty-foot sloop in the British Virgin Islands with his slim wife and four daughters. There he was at the firm-sponsored golf outing and a gallery opening. There he was with his daughters on their mares during a dressage competition and poolside at their country house.

Park's job was to buy small companies, usually community banks or family-owned insurance companies, and restructure them, using borrowed money, in anticipation of a sale or an initial public offering.

It was my job, with Louis's help, to surreptitiously collect any evidence Park had engaged in any wrongdoing at PSP's expense.

Park's bookcase held dozens of crystalline plaques emblazoned with the logos of companies he owned and ran and where he held board seats so he could manage them more directly. This kind of investing—involving the development of undervalued companies owned and run by an umbrella firm whose shares are not traded on a stock exchange—is called private equity.

Louis followed his protocol. After walking the floor to make sure it was empty, he entered Park's office and scanned "high, medium, and low," as he says—above the desk, around the desk, and under the desk—to assess the work space and to find any passwords that were scribbled on Post-it notes.

He located Park's computer under the desk and took photographs with his phone of the position of the keyboard, the wires running across the desk, the angle of the tower, the chair's proximity to the desk. He captured the exact position of everything—stacked books, strewn pens, framed photographs, a stapler—so that he could reconstruct it when he was finished. Then he settled onto the floor, opened the Pelican case, and considered its contents:

- Four two-terabyte Seagate hard drives
- Boot disks
- Two Tableau cloners
- Two write blockers
- Two laptops—one Mac, one PC
- Laptop chargers
- Extra copies of FTK imaging forensic software
- Two iSCSI-to-USB cables
- Two USB-to-USB cables
- Two crossover cables
- A power strip
- An extension cord
- A phone charger
- A flashlight
- Extra flashlight batteries
- A role of electrical tape
- One Phillips-head screwdriver
- One flathead screwdriver
- A digital camera
- A camera charger
- Two chain-of-custody forms

- Three black ballpoint pens
- Two legal pads
- Two bottles of water
- An empty Altoids tin
- Four White Chocolate Macadamia Nut Clif Bars

Louis was a computer forensics technician. I hired him to help me investigate Park to figure out if Park had been stealing information from PSP or using PSP resources without permission or if he had failed to disclose any conflicts with PSP partners, among other possible bad acts.

In my role co-running a private intelligence firm, I often hire specialists with unique skills—forensic accountants, surveillance experts, Mandarin speakers, software developers, and people who excel at finding elusive digital data. I am like a general contractor with a dependable crew of subcontractors—masons, electricians, plumbers, carpenters—to help me build a town house.

Louis—gangly, late thirties, gregarious—began his career in the military, breaking enemy codes and finding moles embedded in U.S. forces or computer systems, then spent ten years working for large public companies (telecoms, banks, health-care providers) in their internal audit departments.

He reported sometimes to the CEOs, sometimes to internal lawyers, and sometimes to a board's audit committees, which review financial statements and oversee regulatory compliance and disclosure.

Louis's job at his former employers was to root out employees acting in bad faith. Sometimes that meant investigating suspicious activity and whistle-blower claims, and sometimes that meant indiscriminate snooping: a kind of corporate stop and frisk. Except the employees never knew they were being stopped and frisked.

Louis had access to emails, servers, and internet traffic so he could eavesdrop on everyone's digital activity. He knew about the porn and the affairs and the gossip, of course, but sometimes he unearthed real crimes: drug dealing, theft of trade secrets, expense account fraud. He got so good at this kind of work for various companies that a friend, a former FBI agent who was setting up his own investigations firm, poached him. Louis became a private detective.

He fit no gumshoe archetype. Louis's skill as a PI was his technical savvy. He had no time for the narrative daisy chains that investigators patiently weave. In fact, he did not even really know why he was investigating Park. It didn't matter. Louis's job was to make a perfect image of Park's hard drive.

Louis wiggled Park's mouse to see if the computer had been left on. (I watched quietly and took notes.) It was on but sleeping.

A live computer's drive can be imaged, or copied, by inserting CDs or thumb drives loaded with software that instruct a source drive to send its data through a USB cable to a destination drive.

But forensic technicians often prefer to remove hard drives before copying them. There are fewer risks.

For instance, plugging a USB plug into a machine leaves a

footprint that could be discovered. And there are advantages to imaging a live drive that has been removed.

If you copy data off a computer that is on, you can also copy its RAM, the temporary storage that exists only when a computer is running. RAM, which disappears when a computer is off, is where the operating system lives, and it's also where malware, viruses, and hacking tools live.

Louis decided to power down Park's computer. (Park's RAM was not a priority.) He suspected Park would notice on Monday morning that his computer had been tampered with. But Louis had a plan.

Computer forensics investigators "image" drives so their data can be preserved and later searched for clues—incriminating emails, illegally downloaded client lists, and websites, forbidden or otherwise.

There are many stories of public figures who unwittingly recorded their frauds.

In 2002, the Citigroup equities analyst Jack Grubman admitted that he had upgraded his rating of AT&T stock in return for Citigroup's CEO Sandy Weill's help getting Grubman's kids into an elite preschool. In an email to a friend, Grubman wrote, "You know everyone thinks I upgraded [AT&T] to get lead for [AT&T Wireless]. Nope. I used Sandy to get my kids in 92nd ST Y pre-school (which is harder than Harvard) and Sandy needed [C. Michael] Armstrong's vote on our board to nuke Reed in showdown. Once coast was clear for both of us (ie Sandy clear victor and my kids confirmed) I went back to my normal negative self on [AT&T]. Armstrong never knew that we both (Sandy and I) played him like a fiddle."

(Despite the emails, Weill and Grubman denied a quid pro quo. Neither man was charged criminally.)

Creating a hard drive's "image" is different from copying and pasting files or dragging and dropping folders.

A forensic image is an exact copy of every shred of data that lives on a computer at the time it is imaged. This includes deleted documents, web browser history, instant messages, data from the unused, or "unallocated," space, and footprints of system activity, such as printing and the dates and times of when exterior devices like thumb drives were plugged into the computer.

Making an image allows investigators to find and "undelete" a farrago of data scraps (Word documents, MP4s, JPEGs, PDFs, texts), carve bits of information out of the crevices of the device, and discover who logged in to and out of the computer and what software was installed or uninstalled. It is even possible to pull data from previous operating systems.

Imaging computers is a crucial, and growing, investigative technique, largely because it allows detectives to recover information thought to be gone forever. And it's legal as long as the hardware and software being searched are owned and controlled by the client. It's the cyber version of going through someone's trash and recycling.

There are, essentially, three ways to "erase" data from a computer: reformatting, deleting, and wiping. A personal computer is constructed like a book with a table of contents. Files are arranged in folders that are structured logically.

Reformatting a drive is like stripping out a single chapter heading from the table of contents without stripping out the chapter itself.

Deleting a drive is akin to erasing the entire table of contents but leaving the body of the book untouched.

Wiping is the most effective way to permanently remove data. This process overwrites every letter, word, sentence, paragraph, and chapter of the entire table of contents and the book.

Private eyes also make forensic images because this technique includes "hashing," in which every file on a source drive, as it is extracted and stored, is assigned a unique code, like a digital fingerprint. Hashing allows investigators to verify that the data we collect is legitimate and unaltered.

If a GIF file is dragged and dropped into a thumb drive, the file's metadata may be altered. If that same GIF is imaged properly in a forensic collection, its properties remain unchanged, like an extinct specimen preserved in a jar of formaldehyde.

Hashing also allows us to create digital timelines (sometimes with millions of entries) that can be mined for clues. And we use hashing to protect ourselves during any testimony we provide during legal proceedings, because it gives us proof of the location and contents of data.

Louis reached around the back of Park's computer and gently tugged at the wires connected to the monitor, the keyboard, the mouse, and the electrical socket to help himself identify wires he would unplug. He removed the books piled on top of the computer, pulled the machine partially out, turned it at an angle, and took photographs of the wires splaying out the back.

Then he switched the power button off, unplugged the wires that didn't give enough slack to allow him to pull the tower out from under the desk, and took out his Phillips-head screwdriver. Louis unscrewed the hard drive and removed it.

He took his Altoids tin out of the Pelican case, placed the screws in it, and snapped the lid shut. He removed from the Pelican case a blank hard drive, data cables and power cords, his laptop, and the write blocker, which prevents any data from being installed on the destination drive. It is not difficult to accidentally install a destination drive's data onto a source drive.

Louis plugged Park's drive into the write blocker, the write blocker into his laptop, and his laptop into the destination drive. Then he pushed a button on the write blocker that began the process of mirroring all the data from Park's computer.

He verified that the transfer was uncorrupted, ate a Clif Bar, found the office kitchen, made himself a Nespresso latte, and waited—for five hours—for the computer to do its work.

*C*lients who fear they have been victims of fraud (and hire lawyers and private eyes to corroborate their suspicions) must sometimes be taught how painstakingly slow it can be to collect and analyze digital information.

The computer forensics field is still in its infancy. Only a decade ago, its tools were not widely available to the private sector. During most investigations like the PSP probe into Park, Louis is first tasked with copying a single drive, but by the end of the project he has cleaved data from multiple hard drives that run on different operating systems, and from various models of smartphones running dozens of apps, as well as from thumb drives and cloud servers.

(My experience is that clients, especially those who run large organizations, rarely know what hardware, software, and data they own, or what is needed during a computer forensics operation.)

All of this data is stored and processed differently, and the tools Louis carries are made by companies like Tableau and Cellebrite that scramble to keep up with rapid innovations in hardware manufacturing.

Once data is imaged, it must be processed, which means uploaded onto a server, then preserved (to protect it as evidence, in case it's needed for a legal proceeding) and pushed into an analytical tool, such as FTK or EnCase, which makes it possible to search. It's the sexless side of a sexy trade.

American courts are still debating how much privacy to grant employees when they use corporate email accounts, computers, and other digital systems controlled by their employers. Many companies require workers to sign contracts laced with legalese, stipulating they have no right of personal privacy "in any matter stored in, created on, received from, or sent through or over" an employer's system, according to one such document I have seen.

Sometimes employers will claim this also covers employees' personal email accounts, like Gmail, or cloud storage accounts, like Box.com.

Companies often reserve the right to review, monitor, access, retrieve, and delete any information stored in, created on, received from, or sent through the employer's system, for any reason, without the permission of any system user and without notice.

PSP's general counsel, relying on the company's outside lawyers' advice, had determined that Park had no expectation of privacy when using his PSP computer, because PSP owned the hardware and suspected Park was engaged in some kind of wrongdoing.

What Louis never learned (and would not have cared to know, because he was hired solely for his ability to copy and preserve digital data) was that Park was planning to secretly steal money from his employer by orchestrating deals that generated huge profits for him.

To accomplish this, he had forged documents that laid out the terms of PSP's planned acquisition of Nabok, a chain of insurance brokers in Texas. Terabytes of paperwork Park helped draft formalized a private equity transaction—representations and warranties, disclosures, a purchase and sale agreement, the rights of investors—that hid Park's connection to the company: his cousin co-founded Nabok, and both she and Park stood to become wealthy if PSP acquired it. Park planned to, essentially, pay himself with PSP's and its investors' funds to buy Nabok.

In a private equity transaction, the acquiring firm itself does not often have a direct relationship with the portfolio company. The firm forms a limited partnership or a limited liability entity, sometimes called a "Newco" if it's undefined, and sets up a fund to buy companies.

Then the firm raises money from investors, such as pension funds, hedge funds, and high-net-worth individuals. The private equity firm is the general partner of the fund and directs its investments, and it has a duty to act in the best interest of the fund. Park failed to do this.

When Louis imaged Park's computer, he unknowingly scraped out of its deleted space documents Park altered to conceal his conflict, which we discovered when we reviewed the imaged files.

By the time Louis screwed Park's hard drive back into the tower and faithfully replaced the items on Park's desk, the sun was reflecting off the midtown towers. Louis had verified that the data he collected was a perfect mirror of Park's computer, and then he covered his tracks—better than Park would.

It feels as if every day we are bombarded with stories about how our personal digital information is compromised. Banks, government agencies, and Silicon Valley firms are constantly under attack by gadflies, spies, and routine criminals alike.

Park could have taken some simple steps to avoid detection. Because he used his work computer to carry out his crimes, he was easily detectable; none of us should use our employers' systems for personal use (especially if it's criminal), because we have no expectation of privacy.

Private investigators are often called on by companies to help them not only uncover wrongdoing but design protocols to prevent it in the future. We sometimes help clients develop systems to audit their digital assets so that crimes like Park's are detected early and can be stopped.

Building defenses into a company's systems is not always easy, because the systems that are infiltrated by outsiders, unlike Park, are vulnerable not necessarily because of the sophistication of the hacker or her tools but because of the outdated and poorly defended infrastructure on which so many of us still depend.

Tarah Wheeler, an information security researcher and political scientist, is articulate on this point. "The nature of cyberwarfare is that it is asymmetric," she wrote in *Foreign Policy*. "Single combatants can find and exploit small holes in the massive defenses of countries and country-sized companies. It won't be cutting-edge cyberattacks that cause the much-feared cyber–Pearl Harbor in the United States or elsewhere. Instead, it will likely be mundane strikes against industrial control systems, transportation networks, and health care providers—because their infrastructure is out of date, poorly maintained, illunderstood, and often unpatchable."

The lesson is that we should update electronic defense networks, software, and hardware often.

The average citizen, too, fails to take even the most basic

precautions to protect her own digital data. This involves installing antivirus software, using password managers that create and store encrypted passwords online, encrypting hard drives, backing up data to both cloud services and external drives, routinely checking credit reports (to check for evidence of identity theft), devising intentionally incorrect answers to website security questions, and employing two-factor authentication.

*L*ouis tracked down Sally, who had stayed in the office all night with us and was nodding off in her cubicle. She directed him to a supply closet near the Nespresso machine where PSP's electric circuit breakers were.

He cleaned the coffeemaker so as not to arouse suspicion, then threw the switches so the system would reset and his tracks would be covered. The lights were out for less than ten seconds. Before they flickered back on, Louis watched the monitors and clock radios and digital picture frames blinking and scrolling in the darkness. He clipped his Pelican case shut, ate another Clif Bar, and went home.

*D*espite my efforts, I lost touch with PSP for a few months. (I always try to cultivate relationships with clients, especially those for whom we crack cases.) But I monitored social media and set up automated news alerts naming the company and Park hoping to learn what PSP did with our findings and what happened to Park. On LinkedIn, I saw that a few months later Park joined a competitor. I also found a PSP press release announcing the hiring of his replacement. I later learned PSP fired Park, but did not pass on our evidence to a prosecutor or regulator. Apparently, PSP feared the negative publicity an indictment would generate. Park got away with it.

3.

DUE DILIGENCE

From a distance, the property looked like the campus of an underfunded state college. But when Joelle Masters and I pulled onto the grounds, it was clear this was no school: the massive Slinkies of concertina wire, the armed guards, and the double-fence perimeter were impossible to misinterpret.

I parked the rental car, and as we walked into the main entrance of the prison, a network of industrial, one-story concrete buildings linked by concrete and steel passageways, we crossed over a rubber floor mat emblazoned with the seal of the Federal Bureau of Prisons.

On one wall hung a list of sartorial restrictions for visitors: no miniskirts, tank tops, khaki, camouflage, and other clothing items that either arouse or confuse (federal prisoners are clad in khaki). It was the summer of 2015.

We signed the visitors' log and checked boxes on forms confirming we were not bringing in weapons, did not have Ebola, had not recently traveled to West Africa, and had not come into contact with anyone who had recently traveled to West Africa.

We surrendered our driver's licenses, passed through a metal detector, and walked through two bank-vault-thick glass doors inlaid with four-inch-thick crosshatches of steel beams.

On the other side, a security guard stamped the backs of our right hands with invisible ink, and as we held up our hands to an infrared light, purple block letters appeared on them: "BOP." Bureau of Prisons.

The guard smiled. "That's your get-out-of-jail-free card," he said.

We were escorted through a courtyard, where inmates were tending to clusters of shrubbery, and into the visiting room.

While we waited for our host, Mark Seffer, a Connecticut financier who was serving twelve years for defrauding investors in his hedge fund out of $140 million, Masters and I awkwardly arranged some folding chairs in a rough semicircle.

To prepare myself to meet Seffer, the previous evening I spent some time on Prison Talk, a website "conceived in a prison cell, designed in a halfway house, and funded by donations from families of ex-offenders," and learned that this facility (which I cannot identify at my client's request) had coin-operated vending machines. I removed from my suit pocket a Ziploc bag containing dimes and quarters. (You cannot give an inmate money, but you can buy her food from vending machines.)

It was the most creative, and allowable, gift I could think of to bring the man I hoped would tell us whether a former employee of his had participated in Seffer's crimes.

On the wall of the waiting room were colorful quilts sewn by inmates. I remembered the postings for "humane" events on the BOP website: "Mother-Child Night of Hopes" and "Mommy and Me Tea."

Then Seffer—in his early seventies, balding, with an unimposing build and a hardened expression—appeared, clad in khaki.

I visited Seffer to ask him about the habits, conduct, movements, affiliations, associations, reputation, and character of David Dasha, a portfolio manager who worked for Seffer before the fund collapsed under an avalanche of investor redemptions (a euphemism for withdrawals) and lawsuits alleging fraud, which led to the criminal conviction of Seffer and some fines by the SEC, which had filed a companion civil case when Seffer was indicted. In other words, I was doing a background check on Dasha. (Seffer and Dasha are not their real names.)

My client was the director of a private equity firm that was considering a $25 million investment to help Dasha expand a fashion company he now owned. She wanted to make sure she was aware of any black marks in Dasha's past before she announced an investment. She agreed to pay me $15,000 to cover research and a handful of interviews, mostly to learn whether Dasha had any responsibility for Seffer's bad acts.

I had worked for this client on other projects, and she was not risk averse. During one investigation, I found that the incoming CFO of a biotech firm she wanted to buy had been arrested for assaulting the boyfriend of a mistress. This was a deal killer, I thought. Surely this fact (which we discovered through interviews with former employees, because there was no public documentation of the incident) would persuade her to reconsider her investment.

She had a different calculation. Although the CFO's extracurricular activities exposed a character flaw, she said, they were irrelevant to his financial acumen and ability to generate value for shareholders. Indeed, his former colleagues, to a man, praised his work ethic, intelligence, and grasp of complex transactions.

I have learned not to predict how clients will react to what I find for them. Investigators are fact gatherers and analysts, not mind readers or legal advisers. We look for hidden or undigested information, patterns, markers. It is my job to find people with valuable information and help them to un-fog their memories, tease out their recollections, talk about things they are often unwilling to discuss. We are not usually paid for our opinions.

Because so many documents—property records, civil court filings, criminal proceedings, and more—in the United States are so widely available to a private investigator, most background checks—which range in price from $500 to $15,000, depending on the reputation, sophistication, and reach of the investigator; the complexity of the project; and the means of the client—consist primarily of database research and collecting filings from court clerks. This research, in the United States, is very comprehensive. We can easily find criminal convictions, civil judgments, property ownership, corporate records, major debts, and more.

We began our workup on Dasha by tracking down and culling through documents to find controversies in his past: criminal records he might have had, civil lawsuits he might have been involved in, exorbitant and recent debts he had accrued, and professional censures by securities regulators. This might include domestic abuse, drunk driving, or wire fraud convictions; litigation over contractual disputes with partners or investors; personal bankruptcies or failures to pay taxes; disciplinary actions by the U.S. Commodity Futures Trading Commission or the U.K.'s Financial Services Authority for reaping ill-gotten gains through market manipulation like unauthorized trading. The spectrum for wrongdoing was broad.

I visited Seffer because during our research on Dasha we learned about Seffer's recent conviction. Although Dasha had not been charged with any crimes, I wondered what Seffer knew about Dasha. The client agreed with my suggestion that it was worth asking him.

Beyond official documentation from Dasha's past, we also looked for negative media reports and embarrassing postings on social media. We went beyond simply entering names in databases and poked around the edges of his life. We looked to see if handles he used for email accounts and as usernames on music streaming and gaming sites were also used to peruse prostitution sites or bad-mouth enemies on LISTSERV.

We also searched for undisclosed businesses he might have had to see if he had other sources of income. And we worked to verify his education and other credentials—a BA in English literature and a CPA license. This credential vetting is among the most crucial elements of a background check, because résumé fudging may suggest dishonesty or worse.

*O*n May 3, 2012, Daniel Loeb, who runs a hedge fund called Third Point, sent a letter to members of the board of Yahoo informing them that the résumé of their CEO, Scott Thompson, was, to put it gently, inaccurate.

Third Point was the largest outside shareholder of Yahoo and was locked in a bitter proxy contest for control of Yahoo's board. A proxy contest occurs when certain shareholders of a public company agitate for some kind of change, such as a sale of assets, a debt reduction, or a management reshuffling, and work to persuade other shareholders to join their cause. Change flows through the board, because the board controls the company.

Typically, an "activist" investor, as these shareholders are called (as opposed to a "passive" investor who buys stock, holds it, and waits for positive returns), seeks to persuade other shareholders to vote out certain directors and elect a new slate that is sympathetic to the activist's aims. The activist seeks to gather these "proxies" to vote his way during the annual shareholders' meeting.

Loeb is an activist whose reputation was built, in part, on his knack for publicly, aggressively, and wittily shaming his targets. In his letter to Yahoo, he reported that Thompson, who had been CEO for only four months, did not have the academic credentials that official biographies of him showed, specifically that he had only one degree from Stonehill College, not two: accounting, but not computer science.

"Upon recognizing this discrepancy," Loeb's letter reads, "Third Point initially assumed that the documents we had reviewed were incorrect and the representations in Yahoo's public filings were accurate. However, we were then informed by Stonehill College that Mr. Thompson did indeed graduate with a degree in accounting only. Furthermore, Stonehill College informed us that it did not begin awarding computer science degrees until 1983—four years after Mr. Thompson graduated."

Then came some sarcasm: "We inquired whether Mr. Thompson had taken a large number of computer science courses, perhaps allowing him to justify to himself that he had 'earned' such a degree. Instead, we learned that during Mr. Thompson's tenure at Stonehill only one such course was even offered—Intro to Computer Science. Presumably, Mr. Thompson took that course."

Yahoo countered that it had made an "inadvertent error,"

but Third Point demanded Yahoo reveal how it vetted Thompson. Loeb wrote that if Thompson "embellished his academic credentials we think that it 1) undermines his credibility as a technology expert and 2) reflects poorly on the character of the CEO who has been tasked with leading Yahoo at this critical juncture."

Loeb encouraged shareholders to question how Yahoo's search committee, which was responsible for recruiting and vetting a CEO, "could permit the Company to hire a CEO with this discrepancy in the public record."

He added, "We assumed previously that the Committee would have conducted a thorough background check on Mr. Thompson—and even if not thorough, the most basic of such checks would address Mr. Thompson's education and degrees."

If Loeb was right, "that would call into serious question whether the Board failed to exercise appropriate diligence and oversight in one of its most fundamental tasks—identifying and hiring the Chief Executive Officer."

(Loeb's letter also revealed that Patti Hart, the chair of Yahoo's search committee and of its nominating and corporate governance committees, held a bachelor's not in marketing and economics from Illinois State University, as her Yahoo bio stated, but in business administration.)

Loeb surely relished digging up Thompson's fabrication, because it allowed him to argue that Yahoo was "in dire need of a complete corporate governance overhaul." Within days of the board's receiving the letter, *The Wall Street Journal* reported, Thompson had been diagnosed with thyroid cancer, and eleven days after Loeb sent his epistolary rebuke, Thompson resigned.

Loeb himself was elected to the Yahoo board, as were two of his nominees. He successfully lobbied for Marissa Mayer, a Google alumna, to become Yahoo's CEO. About a year later, in July 2013, Loeb and his two board allies resigned from the board and sold about forty million shares of Yahoo stock back to Yahoo itself. *Business Insider* reported on Loeb's success: "After Loeb's board takeover, Yahoo soon announced a deal to sell some of its stake in Alibaba," the Chinese internet company. "This netted Yahoo ~$7 billion in cash at the time. . . . Yahoo used the money to buy Yahoo stock. Yahoo's stock price soared," and "Loeb bought his 5% stake at ~$13–$15 per share. Yahoo is now trading at ~$28. Third Point made about $1 billion on the whole deal." It's not clear that this sequence of events was Loeb's ultimate goal, but his active engagement with Yahoo and his investigation of its executives were indeed designed to generate profits.

*I*nvestors are not the only ones who unmask résumé padders. In February 2006, David Edmondson resigned as CEO of RadioShack after he admitted to the *Fort Worth Star-Telegram* that he did not have the two degrees that appeared on the résumé he submitted to RadioShack when he joined the company in 1994. Under questioning, he told the newspaper that he had a ThG, a theology degree, from the Heartland Baptist Bible College, but the paper discovered the college's records

showed Edmonson attended for two semesters and never graduated.

In 2001, Mount Holyoke College suspended the historian
Joseph Ellis for a year without pay after he told students he
fought in the Vietnam War. (He did not.)

In 2007, Marilee Jones, MIT's dean of admissions, resigned
after the school learned that she had lied about having two
degrees from Rensselaer Polytechnic Institute and one degree
from Albany Medical College. Shortly before the revelation,
Jones was interviewed by *The Wall Street Journal* for an article
about how colleges check the credentials of student applicants
and said, "The way the whole college application system is set
up now, it really does encourage cheating and lying."

This brand of fakery is not confined to the United States.
In 2007, Alison Ryan was ousted as public relations chief for
Manchester United after the soccer club discovered she did not
earn a first-class degree in history from Cambridge, was banned
from practicing law, and forged a reference from a lecturer at
Manchester Metropolitan University who did not exist.

The same year, *The Washington Post* reported that an "epi-
demic of phony academic credentials" broke out in South Korea, "a nation where calibrations of human worth are obsessively
tied to college achievement." The article continues, "Admired
performers, beloved media personalities, assorted academicians
and a revered Buddhist monk have been exposed as long-time
résumé inflaters."

Prosecutors in Korea began a nationwide investigation into
"fabricated degrees, plagiarized doctoral theses and forged test

certificates." Among the "most unrepentant" was Shin Jeong-ah, an art history professor at Seoul's Dongguk University and "the youngest-ever artistic director of a major arts festival." Her résumé showed she had a doctoral degree from Yale and bachelor's and master's degrees from the University of Kansas, but Yale told the *Associated Press* she never attended the school, and although she did attend the University of Kansas, she never earned a degree.

*B*iographies of Dasha posted online by his employers and regulators as well as aggregators showed his bachelor's degree in English was earned in 2003. But when we asked the National Student Clearinghouse, a nonprofit organization that has secured a virtual monopoly on the degree-verification business in the United States, to confirm this, it found no supporting documentation. We called the college's registrar and explained our dilemma, but no one there would release any information to a third party.

Why did some bios of Dasha posted online show he was a college graduate, but the school's records failed to verify this? Was Dasha a liar or the unwitting victim of a typo? I prepared to tell my client that Dasha did not have the degree he claimed.

Seffer was not shackled when he met with us. He wore pants, a short-sleeve, button-down shirt, white tube socks pulled calf-high, black sneakers fastened with Velcro straps, and frameless eyeglasses. The guard who accompanied him announced casually he would return in two hours. We were alone with the felon.

Masters and I introduced ourselves to Seffer, who shook our hands, chose a seat in the U-shaped collection of the few chairs we had arranged, interlocked his fingers on his lap, and said, "How can I help you?"

I explained that during a background check on Dasha we learned about Seffer's conviction. I asked if he would discuss his former employee. He agreed. But before we discussed anything in detail, I made an effort to win his respect and prove my credibility by telling him how I approach my work and offering to answer any questions he had for me beforehand. We engaged in small talk for fifteen minutes. He told me about life on the inside. He denied wrongdoing. He was a scapegoat, he said, the victim of an overzealous prosecutor with political ambitions. He would be released soon. He was not like the other prisoners. He was different. He was smarter. He was doing legal research because most of his friends and family had abandoned him. But they were wrong. He'd show them. As soon as he was out.

I didn't argue. I nodded approvingly, then began my inquiry.

"Let's start at the beginning. How did Dasha come to work at your firm?"

"I don't remember, but I imagine our human resources department used a headhunter."

"Was he a talented investor and manager? How would you assess his work and reputation?"

"Dasha is smart, headstrong, ambitious. But I didn't really know him well."

"Really?"

"No. I delegated oversight of portfolio managers to my chief investment strategist." He paused. "It's funny—we almost didn't hire him."

"Why?"

"He was sloppy."

"How?"

Seffer smiled as if remembering something fondly. "There was a typo on his résumé. Who misspells his own name?"

*M*any states require private detectives to pass a written test as part of the license-application process. The New York exam consists of a hundred multiple-choice questions covering the PI statute, basic investigative techniques, evidence gathering, and elementary management theory.

There are also terms the applicant has to define—"alibi," "forgery," "subpoena *duces tecum*," "fiduciary"—as well as some more colorful terms: "jostling" ("unnecessary crowding against a person with the intent to place a hand in the proximity of such person's pocket, pocketbook or handbag"), "malingering" ("the act of feigning or faking an illness"), and "rough shadowing" ("shadowing which is done in such a manner as to cause the subject discomfort or annoyance, or interferes with him in any way").

The exam is offered monthly in about a dozen cities across the state, and in Manhattan the test center, when I took it, was a few blocks south of the Brooklyn Bridge in a building that also holds a field office of the Social Security Administration, an outpost of the American Real Estate Institute, and the New York State Court Officers Academy.

On the second floor of the same building is the New York Department of State's Division of Licensing Services, which oversees private detectives and, according to its websites, the licensure, registration, and regulation of more than thirty other occupations: armored car guard, athlete agent, bail enforcement agent, hearing aid dispenser, home inspector, milk broker, nursery dealer, steamship operator, soil inoculator, and cosmetologist.

Although the PI statute covers bail enforcement agents, watch guards, patrol agencies, and security guards, the private investigator's exam covers issues related only to investigative work. (There are separate exams for the other professions covered by the statute.) When I took the exam, in a fluorescent-lit, windowless room, there were only three other gumshoe wannabes in attendance. The rest of the people in the room were muscle. I had the slightest frame and the fairest skin in the room. One tall, round man wore a baggy black T-shirt with white block lettering on the back: "SECURITY."

Although private eyes are permitted to conduct a broad range of investigative tasks, the questions on the exam cover a narrow band of detective work, such as arson, automobile accidents, insurance fraud, and domestic disputes. The drafters of the exam did not envision assignments that betray the stereotype of the surveillance expert and the process server, who spend their days hanging around courthouses and nights rooting through dumpsters.

There were no questions on my exam covering insider trading, money laundering, tax evasion, counterfeiting, channel stuffing, patent infringement, or offshore shell companies. There are no questions about software for extracting deleted texts from iPhones, uncovering violations of the Foreign Corrupt Practices Act, helping Nasdaq-listed firms defend against activist shareholders during a proxy contest, or finding evidence that a hedge fund is leaking negative news about a company it is shorting—all of which I've come across in my work.

One question covered obtaining plaster casts of tire treads. One considered countersurveillance techniques. One was about analyzing handwriting samples (a "scientific" technique whose credibility has been challenged).

Some states do not require that private detectives take an exam; others require classroom training. While some states have reciprocity, most do not. A California detective may not work in, say, Virginia without first obtaining a Virginia license or working with a licensed Virginia-based private eye. Virginia has limited reciprocity agreements with Florida, Georgia, Louisiana, North Carolina, Oklahoma, and Tennessee, but not New York.

Although I had my New York license—I passed the exam, completed a background check, paid a fee, obtained affidavits from former colleagues testifying to my good moral character and experience, and filled out innumerable forms—I was not permitted to operate in Virginia, where Seffer was imprisoned, without a local chaperone.

This is why I hired Masters, a retired probation officer turned PI with a Virginia license. During our interview with Seffer, Masters barely said a word. Her presence was all I needed. She was my silent protector, my fixer.

*D*avid Desha—not Dasha—was the name on Dasha's résumé. That's what Seffer told me. It was a tweak, not a wholesale reinvention, but it was enough of an alteration to conceal his identity so that Desha could work for Seffer. (I later learned that Seffer's firm did not run background checks on employees and so had no opportunity to unmask Dasha when he was hired.) Seffer's memory was that Dasha had simply not noticed a typo on his own résumé.

Armed with this new information, my colleagues and I went back to our databases and reran the searches. Among our first steps was to verify Desha's college degree. We knew David Dasha had not earned an English degree as his résumé claimed; David Desha had not either. And here was our first breakthrough: Desha did attend the school, but he had dropped out after two years. He hadn't earned a degree at all.

We now expanded our searches. Because of Dasha's role at Seffer's hedge fund, he had to have been a licensed broker or dealer. FINRA rules state that anyone "engaged in the securities business" of a firm, "including salespersons, branch managers, department supervisors, partners, officers and directors," must be registered as a broker or dealer by their firm with FINRA.

We knew this, so we obtained a BrokerCheck report on Desha from the not-for-profit organization Financial Industry Regulatory Authority Inc., or FINRA, which oversees more than 3,500 securities firms with about 625,000 brokers. (We had already searched this source for records naming Dasha.) Its congressional mandate is to examine firms for compliance, foster market transparency, and educate investors.

We also asked the Connecticut Department of Banking's Securities and Business Investments Division for a copy of his record, which is called a CRD (for Central Registration Depository) report. (Seffer's firm was based in Connecticut, and this is where Desha worked.)

A CRD report is gold for a corporate PI (or any consumer, for that matter) who seeks to understand the background of a financial professional. It shows what securities exams a broker has passed or failed to pass, lists her former employment and firms with which she has been registered as a broker, and shows any "disclosure" events such as crimes, lawsuits, or sanctions.

These records are self-disclosed, however, and submitted to FINRA by the employer based on information from some questionnaires and forms the broker fills out. Some brokers are more revealing than others. Someone I once investigated disclosed a quarrel with an Uber driver over a $3 tip. Another failed to disclose he was censured by the SEC for shorting the stock of a company his sister, an investment banker, helped underwrite before its IPO.

Desha's CRD showed that in April 2001 the Securities and Exchange Commission charged him and three other traders at his firm with violating regulations governing the practice of lending money to customers to buy stocks. Desha made dozens of loans worth more than $6 million to customers using an account set up in his name. This credit was provided to customers who could not cover margin calls (demands that investors front money as collateral to cover possible losses).

Desha did not authorize any loans, did not learn the identities of the borrowers, and did not approve the creditworthiness of the borrowers. His customers paid $5,000 in fees for the loans, and those fees were fed into Desha's account. It was theft. Desha was censured, subjected to a cease-and-desist order, instructed to pay a $10,000 penalty, and suspended from any association with any broker or dealer for six months—which is about how long he waited to apply for a job with Seffer.

D esha spent no time in prison. And yet a prison is where we ended up investigating him. The Federal Bureau of Prisons was formed in 1930 to provide "more progressive and humane care" for federal inmates, of which there were 13,000 at the time. Today there are about 150 federal facilities in the United States (some of them privately run) that hold about 165,000 inmates. Eighty percent are American. More than 10 percent are Mexican. Approximately 60 percent are white. Almost 7 percent are female. While nearly half of federal prisoners are removed from society for drug offenses, less than 6 percent are in for white-collar crimes: banking and insurance fraud, counterfeiting, embezzlement, extortion, or bribery, according to BOP data.

Visiting an inmate in a federal prison requires research and planning. I found Seffer by searching the BOP website. His record included his name, the location of his prison, his age, his expected release date, and his register number, a unique identifier the federal government issues inmates to track them.

Convicts are often held in facilities outside their home states, but not necessarily in states with a connection to their crimes. For instance, Bernie Madoff—the former New York money manager who bilked billions from investors in the largest fraud case in U.S. history—spends his days in a medium-security complex in Butner, North Carolina, an hour northwest of Raleigh. Rajat Gupta, the former McKinsey chief and Goldman Sachs director, who was convicted in 2012 of feeding insider information to the hedge fund manager Raj Rajaratnam, was in an "administrative security federal medical center" in Ayer, Massachusetts. Ted Kaczynski, from Illinois, was living in seclusion in Montana when he was arrested for being the Unabomber and today is in a "supermax" facility in Florence, Colorado.

To obtain permission to visit a federal inmate, one must submit an application that includes prior addresses, employment history, and consent to a background check through the FBI's National Crime Information Center. Inmates must then approve visitors' requests for meetings and also to communicate over email through a system called CorrLinks, which is managed by

Advanced Technologies Group, a private company in Iowa that sells software and services to the Federal Bureau of Prisons.

*A*n hour into our meeting with Seffer, he was still sitting erect, prideful, and imperious. Seffer remembered little about Dasha. Neither Masters nor I knew at that point that Dasha had changed his name to cover up his lie about having a college degree and being banned from the securities industry.

Our conversation meandered: politics, commissary food, his career. Seffer yearned to talk as long as his guard, who only once wandered into the room and said nothing, would allow. As our conversation came to a close, he surprised me.

"You have not asked me the question you came here to ask," he said. (I did not record this conversation; I rarely record witness interviews. Instead, I took some notes.)

"What is that?"

"Whether Dasha broke the law."

I planned to ask him that, of course, but did not want to burst through the front door.

"The answer is no. No one at my firm, including me, engaged in criminal activity."

Seffer then began to ramble. He repeated his pleas of innocence and described how one of the prosecutors in his case quit soon after Seffer was indicted, which Seffer interpreted as evidence that the government felt its case was weak.

We wound down our conversation by discussing Seffer's appeal, which had still not been ruled on by the U.S. Court of

Appeals. He was cordial when we shook hands. He watched us walk out and back through security. I never saw him again.

Several weeks later, when we completed our investigation, I wrote a report summarizing our efforts and conclusions. Although Dasha had black marks on his record, he had never been accused of crimes related to Seffer or Seffer's firm. (I spoke with one of the prosecutors who tried Seffer's case, and she confirmed there were no other suspects.)

But Dasha was definitely a liar and had been banned from the securities business. His subterfuge had worked on Seffer, who may die in prison (he was sentenced to eight years).

Because of what I found, my client, the consummate risk taker, declined to invest in Dasha's fashion company. His track record was simply too fractured.

*M*any of my clients decline to invest in prospective transactions as a result of damning information I find for them. The smartest ones understand that it is not only felonies and mob ties that kill deals but seemingly innocuous behavior like résumé padding that has profound consequences.

Board directors of private companies that hope to sell shares to the public are discarded by underwriters, who may recommend their clients delay listing on an exchange.

Executives of auto-parts manufacturers are replaced by their private equity masters, who are not afraid to retool entire management teams.

Would-be city controllers are rejected by government

personnel departments or outraged voters, if the controllers are high-profile allies of elected officials.

Although the capital markets (and the internet) can be unforgiving—companies lose revenue, brands are tarnished, careers are grounded—the transparency that comes from such revelations can be beneficial, even healing, to commerce and society.

Daniel Loeb, for instance, made a fortune off his research and maneuvering around Yahoo's management.

Although the bad news about RadioShack's CEO was one in a series of missteps (after several failed turnaround attempts, it filed for bankruptcy protection in February 2015 after nearly a century in business), the electronics retailer appeared to learn a lesson. It temporarily removed from its website the biographies of its officers and replaced them with a note: "We are currently updating and validating all of the biographical information for each of our senior executives."

Not all companies are so inspired to investigate their senior executives. I learned this from some former British spies who chased an American fugitive to the South of France.

4.

THE FRENCH VET

In the summer of 2011, Michael Mastro, a real estate developer in Washington State, and his wife, Linda, a grade school teacher turned socialite, retired. The couple had amassed considerable wealth and lived opulently. Among their multiple homes, they owned a seven-thousand-square-foot mansion with a pool, a waterfall, and a sauna on Lake Washington in the tony Seattle satellite of Medina. They collected wine and Chihuly glass artwork. They drove a Bentley convertible and a Rolls-Royce. Linda favored Chanel handbags, Louis Vuitton luggage, and Manolo Blahnik shoes and owned countless jewels, including brooches inlaid with diamonds, strands of mixed rose quartz and cloisonné beads, and ivory carved rings.

But Mastro, then in his mid-eighties, had recently suffered a head injury after a sharp fall in his garage, which put him briefly in a coma. Though brain surgery relieved some of the pain that lingered after the coma—and despite vacations in

Italy, Paris, New York, Palm Springs, Switzerland, and Jackson Hole, Wyoming—Mastro decided it was time for a more permanent change. He and Linda—at sixty-two, a generation younger than her husband—packed their bags and, along with their three Shih Tzus, moved to Europe, settling in Veyrier-du-Lac on the banks of Lake Annecy in the Haute-Savoie region of France, about an hour south of Geneva.

They found a house to rent for about 5,000 euros a month. Linda started taking French lessons, and the couple spent their days watching paragliders float over the cerulean lake, which is ringed by snowcapped peaks and postcard-perfect medieval villages with red-tiled roofs. Flea markets line the narrow alleys of the old town. Cyclists favor the roads around the lake, claimed to be Europe's least polluted body of water.

The region has inspired landscapes by Cézanne and poetry by Alphonse de Lamartine. It is where Jean-Jacques Rousseau went as a teenager seeking spiritual renewal. "God is calling you," a priest advised Rousseau, as he writes in *Confessions*. "Go to Annecy."

Mastro's retirement was palliative for other reasons. For nearly two generations, he bought, developed, and sold residential apartment complexes, shopping centers, and office parks, mostly in Washington State, and he lent money to other real estate investors, but his ventures were buttressed by tremendous debt. The economic downturn of 2008 had ravaged the real estate market. Mastro stopped paying his lenders largely because his borrowers stopped paying him. And his debts were growing faster than his income.

Three of Mastro's largest creditors, hoping to stanch the flow, sued in 2009 to compel him to file for bankruptcy. His liabilities were estimated at more than half a billion dollars. The federal bankruptcy court appointed a Seattle lawyer named James Rigby as trustee to recover assets for the creditors who likely feared they would receive pennies for every dollar they had invested with Mastro.

Not long after he was hired, Rigby, who has done trustee work for more than twenty-five years, began to doubt Mastro's credibility. He suspected Mastro was failing to disclose what he had and was illegally evading creditors. Rigby sued Mastro for hiding assets in offshore trusts, among other maneuvers.

When the bankruptcy judge, at Rigby's request, ordered Linda to turn over two diamond rings—one pear cut of 27.8 carats and the other round brilliant cut of 15.93 carats—worth about $1.4 million, the Mastros stalled: the rings were unavailable or lost; returning them was inconvenient; the couple had fallen ill.

It was when these excuses ran out that the Mastros made for the Rhône Alpes. They stuffed suitcases and duffel bags full of jewelry, clothing, and cash, bundled it all into a Range Rover, drove to Toronto, and shipped the car to Lisbon, where they later landed in a private jet paid for by a friend. After making their way through Europe, they chose Annecy as their hideout. (It was, as it had been in Rousseau's case, a priest who had suggested this location to Mastro, according to one media report.)

By then, the Mastros were eluding the courts, irate creditors (nearly two hundred of whom were small investors from western Washington State), confounded friends, and baffled family members, including Gloria Plischke, Mastro's sister, who had entrusted him with her investments. Plischke would soon be living off Social Security and a loan from a reverse mortgage.

The U.S. Marshals Service, whose federal mandate is, in part, to apprehend fugitives, and the Federal Bureau of Investigation joined the chase for the couple. The FBI, an agency of the U.S. Department of Justice, which also oversees bankruptcy proceedings, had already been investigating Mastro for more than a year, since Rigby had unearthed evidence that Mastro had unclean hands (such as moving assets offshore) and shared what he had learned with federal investigators.

Rigby was frustrated by how territorial and uncooperative the various agencies were. "Dysfunctional" is how he described their tangled relationships. The U.S. Attorney's Office in Seattle, which eventually opened a criminal investigation of Mastro, "wanted information from me," he said, "but they wouldn't tell me what day of the week it was."

Rigby had already started to collect that information. One source of his was Daniel Hall, a private detective who ran a nascent British investigations firm called Focus, and Robert Capper, a former MI6 counterintelligence officer who became

Hall's first employee. Hall and Capper were sipping espressos in Hall's London kitchen one Saturday morning in late 2011 when Hall's mobile phone buzzed.

It was an email from an American lawyer who had been retained by Rigby to help track down Mastro's money in Europe. At the time, Hall and Capper had recently left GPW, a British business intelligence firm; Focus was only a month old.

Hall took the case, but he politely disagreed with what he felt was a rigid and narrow assignment: to search for documents, such as property deeds, in Portugal that might hold clues to Mastro's wealth. (Rigby and other American authorities knew Mastro's plane had landed in Lisbon, but no one yet knew he was in Annecy.)

"The client was very process-driven," Hall explained. "They thought in a linear way: 'Get discovery on Mastro's American Express card to try to track his movements. Then spend 10,000 pounds to do database research in Portugal.'" He described this approach as "boring and doomed."

Hall could have scoured Portuguese property records, searched for media reports and internet postings, dug up corporate filings, and combed through court records. And he would have found nothing, written a report detailing his efforts, collected a fee, and moved on to the next case. Instead, Hall proposed a more expansive, expensive, and innovative exploration that involved variegated fieldwork.

Hall—who would later sell Focus to a global litigation investment firm called Burford Capital, which is based in London—is an expert in the field of asset tracing. He specializes

in what lawyers and investors call "judgment enforcement"—
that is, debt collecting. Hall's experience told him to be skeptical of the speculation that Mastro had put down roots in
Portugal.

Hall is tall, boyish, and self-deprecating, yet he projects
gravitas. His success in winning over reluctant witnesses de-
rives, in part, from his willingness to confess his ignorance
when someone is explaining a complex issue to him. He is not
unlike Denzel Washington's character in the film *Philadelphia*.
"Explain it to me like I'm a four-year-old," he'll say.

Soon, the Mastros inadvertently tipped off Rigby, twice,
to their whereabouts—first when Mastro submitted a claim to
his American health insurer for an ear operation he had under-
gone in France, and again when Linda registered their Range
Rover in France. ("I have to do things right," she would later
tell *The Guardian*. "I never even jaywalked.")

The American authorities, including Rigby, enlisted their
French counterparts for help, but prosecutors in Annecy were
overwhelmed with an odd spate of unrelated mysteries including the shooting of a British family and a French cyclist as well
as several drownings in Lake Annecy, normally a model of
tranquility.

But the molasses-like movements of French law enforcement didn't matter much: Hall and Capper were already on
the move. Although Rigby now knew that Mastro was in or
near Annecy, his exact whereabouts were still unknown. Now
that everyone agreed that the Mastros were not in Portugal but

in the Annecy region, the lawyers suggested that Hall and Capper ask people in the area about Mastro and his financial status.

"This was absurd," said Hall. First, the lawyers didn't know exactly where he was, and now they were suggesting a futile tactic. "How often do you tell your neighbors about your secret, illegal self, especially if you're a fugitive?" Hall asked.

Capper agreed, explaining that often his clients try to solve their problems themselves, suggesting how to resolve cases, even though they are paying tens of thousands of dollars for a top investigator to do this work for them.

"The best thing for us is when clients tell us their problems and we come up with solutions," he said. "One thing that was impressed on me during my time in the intelligence services"— where he was also on the counterterrorism desk—"is that you have to think about the real-life situation. It's exactly the same with this. You can approach these assignments very coldly, reading reports, talking about shareholdings, and it's just crap. The world doesn't work like that. You don't wake up one morning and just decide to move all your British Virgin Islands holdings to the Seychelles. You do that for a reason."

Hall and Capper asked themselves what Mastro would be doing with his days, what his hobbies were. They theorized that because Mastro fled in a private car and then on a private plane—and because the Shih Tzus were also missing—he had brought the pets with him.

"Find the dogs," Hall concluded, "and you find the man."

---

*U*ntil the nineteenth century, some countries physically removed from public life citizens who were burdened with debt by locking them in debtors' prisons. The United States had abandoned the practice of jailing debtors by the 1830s. (The British stopped imprisoning insolvents in the 1860s.) Most countries today favor creditors, but in America debtors have considerable rights. If an individual or a company cannot pay its bills, there is a legal right to file for government protection, which effectively permits a debtor to suspend paying its bills while it negotiates with creditors to achieve a settlement of its claims. (In some cases, there is no negotiation, only a distribution of funds based on statutory priorities.)

Under U.S. law, there are a few types, or chapters, of bankruptcy. Among them is Chapter 11, under which a company can reorganize and continue operating, and Chapter 7, under which a company sells its assets. (Mastro was forced into involuntary Chapter 11, which was later converted to a Chapter 7.) At the end of the bankruptcy process, debts are discharged and can no longer be claimed by creditors. Federal bankruptcy laws afford debtors a fresh start—without prison.

In a Chapter 7 bankruptcy, a trustee—often a private attorney who is a member of a vetted panel—is appointed by the Department of Justice, which oversees all bankruptcies. (Until 1979, judges appointed trustees, but that practice was abandoned to avoid cronyism.)

Trustees are granted extraordinary powers to marshal a debtor's assets and distribute them to creditors. Most of their work—untangling who owes what to whom—takes place over phone calls and in meetings, but trustees can also enlist the U.S. Bankruptcy Court by obtaining the power to subpoena documents, such as bank records, and sue to get more information from the debtor or others whom the trustee believes owe money to the creditors.

A lawsuit filed by a trustee is called an adversary proceeding, and it often alleges "fraudulent conveyance," which means the defendant has improperly moved money out of, or received money from, the estate that the trustee determined should be clawed back. Rigby sued Mastro in this way in an effort to get Mastro to turn over assets or, at the very least, turn over information about his assets to Rigby.

Trustees can hire law firms, accountants, and private eyes and deploy armed guards to seize assets; they can hire liquidation companies, engage bounty hunters, and auction off assets such as jewelry, real estate holdings, precious metals, art, furniture, and wine collections.

The most powerful tool available to trustees, however, according to Albert Togut, who has served as a bankruptcy trustee in thousands of cases during the past forty years, is Rule 2004, which allows trustees to compel any party even tangentially linked to a debtor to hand over documents—internal emails, contracts, financial reports, and so on—or be deposed.

"Its scope is much broader and intrusive than civil discovery, and therefore often referred to as a 'fishing expedition,'" Togut said.

Rule 2004 is one reason legal observers often view bankruptcy trustees as omnipotent. "I can't tell you how many times I've been called 'Your Honor,'" said Togut, who was not involved in the Mastro case but spoke with me about bankruptcy law and the role of trustees.

For all their power, bankruptcy lawyers—who act not only as trustees but sometimes also as counsel to debtors and creditors and as mediators—are a lonely bunch. Law students and ambitious young attorneys more often dream of becoming powerhouse civil litigators, theatrical criminal defense attorneys, or high-stakes deal makers. Bankruptcy law rarely makes the glamour cut.

"When I was coming up" in the 1970s, said Togut, "bankruptcy law was the underbelly of the legal profession." Until recently, he explained, many of the large law firms eschewed the field. But today, many—including Sullivan & Cromwell; Kirkland & Ellis; Skadden, Arps, Slate, Meagher & Flom; and Milbank—have thriving restructuring practices. And yet, Togut conceded, "we are still not part of the public consciousness."

A recent exception is Irving Picard, the trustee for the estate of Bernie Madoff's defunct firm, which collapsed in late 2008, when Madoff was arrested. Picard's efforts on behalf of Madoff's victims and creditors, including hiring investigators from firms like FTI Consulting, Duff & Phelps, and K2 Intelligence, through the law firm BakerHostetler, have been widely covered in the media, exposing a broader audience to what bankruptcy trustees actually do. As of this writing, Picard had recovered more than $13 billion, which includes settlements from lawsuits he filed.

The Mastro saga was also covered widely by the media—in part because it was the largest personal bankruptcy in Washington State history, but also because it was great copy: the Mastros were rich and on the lam.

*T*his is going to sound like a crazy question," Capper said, holding up a photograph of Mastro. He was speaking to a French veterinarian near Annecy. The picture was of a man with a prominent nose and a tuft of thick white hair. "Do you recognize this man? This is Mike Mastro. He's an international fugitive."

Mastro's dogs were old and "would have been preened to within an inch of their lives," Capper figured, given Mastro's extravagant lifestyle and fondness for his pets. No vet would forget them—or the Mastros, either, an elderly American couple who did not speak French and were both known to favor tracksuits.

"Yes," said the vet. "I do."

Capper, who speaks some French but is not fluent, had not used traditional spy games to track down the Mastros' vet. Instead, he'd googled veterinarians in the area, picked the poshest one, and stopped in unannounced.

The investigator dropped a dose of dry British wit and

"cheery persuasion," he told me, and after a time the vet handed him Mastro's file, which included payment slips and credit card receipts.

"In a case like this, you work the distinctive details," he told me—"the decrepit dogs." There is no eureka moment. "All you can do is work through it in the order you're given."

Throughout the case, Capper traveled back and forth to Annecy for two months deploying his affability and charm all over town. He spoke, usually in English but occasionally in broken French, with court clerks, waitresses, lawyers, bartenders, translators, police officers, federal investigators, real estate agents, landowners, court officers, barbers, detectives, security guards, neighbors, and bankers.

Capper's sharpest tool was not a hidden camera or a listening device or a wiretap or a satellite image or a disguise or a bribe. It was his charisma. He could get people to talk to him.

Armed with as much information as he could collect, Capper walked up and down the cobblestone streets of nearby villages in near-freezing weather through newly fallen snow looking for the Mastros' residence. He read the names on mailboxes, gates, and front doors. Capper was not given to self-doubt, but even he was beginning to believe this was a fool's errand—until one day he came across a mailbox in front of a house with the name Mastro plainly stenciled on the side.

To Capper's surprise, the Mastros had not been living in seclusion. Capper snapped a few photographs and sent them to the client. Finding the couple, however, was not really Capper's job; it was finding their money. In fact, by the time Capper found their home, they had already been arrested by French authorities, at the request of the Americans, and were in a holding cell across town.

It didn't much matter where they were, explained Capper. "We needed to know where the hell their stuff was," he said. "Did they have safety-deposit boxes? Where were the bank accounts? Where were the iPads with the passwords on them?"

He continued, "It's the difference between criminal and civil investigations. If somebody steals your handbag, do you want him arrested or do you want your handbag back?"

The arrest proved helpful. Once the word got out, everyone in town began to gossip about the American prisoners. And it didn't take much prodding to persuade locals to speak.

On Monday morning, Capper went to see the Annecy police chief. He was careful not to behave like a journalist— "speaking to everyone, like a leech, bleeding information out of people. You get results by knowing when to talk and most importantly knowing when to not talk. You can always step back, but once you speak you can't unspeak."

At police headquarters, Capper saw throngs of "noisy, urgent, shouty" reporters, so he backed off, found a café, ordered a cup of tea, and waited. When the journalists left the police station, he requested an audience with the chief, explaining he was there in an official capacity, representing the Mastro trustee. The chief invited him into his office, and Capper offered information, such as details on Mastro's American life, instead of requesting it.

"I was going to him as an information equal rather than

fishing for stuff, sitting there with a Dictaphone." During that first meeting, Capper asked the chief few questions. "I prefer to let other people do the talking. If I have to, I sit for a minute. I don't fill the silence."

Over the next few weeks, Capper met with the police chief half a dozen times. They developed a rapport. Most of what the chief told Capper was not valuable, except for one tidbit: the Mastros had previously rented a different villa nearby. This gave Capper another place to look for evidence.

A week later, Capper accompanied a French bailiff to the Mastros' most recent apartment to audit their belongings. Capper was legally Rigby's agent, but he could not seize anything; criminal investigators had the right-of-way.

Throughout the investigation, Capper had to navigate, and in some cases referee, the squabbles among law enforcement agencies, regulators, and courts from two countries. Rigby filed suit in France to establish himself as a trustee there so he would have priority to seize assets, but, he said, the FBI tried to overrule him, claiming it needed to keep assets as evidence in its criminal investigation. Rigby said the French police were often willing to let Hall and Capper "tag along"; the FBI was not.

Capper was awed by the bounty investigators found in the Mastros' house. Bags of Cartier watches, necklaces, suitcases of unworn Dior suits, furs, were stacked up, spilling over, tucked in drawers. "Hundreds of thousands of pounds of jewelry and clothing that had never been worn, never been used, just in a back bedroom, thrown into bags, piled high in closets," said

Capper, who laughed when recalling how ill-prepared the local French bailiff had been.

The court officer had driven a small car anticipating a few items, maybe a couple of pieces of jewelry and a fur coat. "He needed to call in backup," Capper said.

The French police would eventually consign these goods to various American authorities, including Rigby, who eventually sold items to repay Mastro's creditors.

While ambling through the apartment, Capper noticed a business card on a shelf. A portion was torn off. It was for a pawnshop in Annecy. To anybody else, the card might have been a stray piece of litter, but Capper guessed that the police had pulled it from Mastro's wallet when they frisked and arrested him.

"Three weeks later, when I was driving to one of the Mastros' court appearances," Capper recalled, "the penny dropped." He realized this was where Mastro would have gone to sell expensive belongings. He turned his car around and drove directly to the shop, which was in a rough part of town in a region not known for rough parts of town.

At first, the owner was reluctant to divulge anything, but after some friendly persistence Capper was handed Mastro's file.

Capper had also learned that Mastro was a fastidious note keeper. While rummaging through his apartment with the bailiff, he had found meticulous ledgers. "Mastro made his own bank statements. He recorded what he bought, what things cost, what he needed, what his balance was."

This helped Capper to connect the many disparate dots. "Every few weeks, Mastro would go down to the shop with some jewelry, maybe a diamond ring, some ornament, maybe some silver candlesticks, a gold necklace, and pawn them for gold or cash. If he was paid with a check, the check would go into a bank account."

This discovery offered a breakthrough. Capper felt that until then he had been casting frantically about for clues. But now he began to understand the narrative of Mastro's secret life.

"Our client, journalists, locals, and others were all floating conspiracy theories about where the Mastros kept their money—in the Bahamas, in secret accounts in the B.V.I.'s. But it was much simpler. Mastro was not accessing a secret, offshore pot for this life in France. Watches went into the pawnshop, and 5,000 euros in gold or cash would come out and go into their bank account," said Capper, who compared the shop's records with Mastro's ledgers and found that they lined up neatly.

"The rent would come out. Food would come out. They bought petrol. They had cash to spend. And Mastro accounted for every penny, whether it was a 1-euro toll or 57 euros for lunch."

The Mastros were careful not to structure transactions that would trigger suspicious activity reports, designed to identify patterns in the global banking system that would suggest money laundering or other fraudulent activity.

*I*n the United States, the Financial Crimes Enforcement Network, a unit of the Department of the Treasury, regulates money-laundering investigations under the Bank Secrecy Act. Financial institutions must keep records of transactions that law enforcement agencies use in criminal, tax, and regulatory probes. Some alerts are activated by monetary thresholds, such as cash transactions exceeding $10,000 during a single day by or on behalf of the same person.

There are dozens of related rules. One, for example, requires certain institutions to track and record money orders or traveler's checks issued for between $3,000 and $10,000.

Among Capper's cleverest discoveries was what financial institution the Mastros had used. "There were so many authorities who were coming after these guys, and no one could figure out where they banked."

As always with Capper's investigation, tact was key. When he approached the Mastros' landlord, who operated a series of luxury villas in the region, he knew that real estate was the landlord's livelihood and that the arrival of police at his door would surely be disruptive. "He wanted to keep a lid on the Mastro drama," said Capper. "It was not in his interest to have lots of noise."

The landlord felt personally slighted, too. The Mastros had been living within his compound. They spent Christmas together. They had become friends. "He thought the Mastros were a sweet old American couple doing some prolonged travel around Europe," explained Capper.

When engaging people who have been close to fraudsters, "you're walking on eggshells from the start," Capper went on. "It's a live subject. The Mastros had been arrested, were awaiting trial for international fraud, and facing deportation. People are aware there is a legal process they don't want to obstruct. Being able to carefully tease out enough detail without making it feel like a 'smash and grab' job was crucial. From the start, it was planned as a series of meetings. We met four or five times over three weeks. Some meetings lasted five minutes, some an hour."

During one conversation, the landlord confided in Capper that when the Mastros arrived in Annecy, they told him they did not speak French and needed help setting up a bank account. They offered to have their rent wired directly if the landlord would set up an account for them as a sub-account of his own. He agreed. It was not a far-fetched idea, because it is so difficult for foreigners, especially transitory ones, to open accounts in France.

"The Mastros could take out cash from ATMs and transfer money," said Capper. "This was a small town of a few thousand residents, and it was all easily arranged because the landlord was good friends with the bank manager."

Capper, despite his ability to cajole information from others, was astounded that people were willing to tell him about a stranger's banking arrangements, simply because he asked.

He spoke with the bank manager at Société Générale while

sipping coffee in the manager's office in Annecy, and the manager confirmed much of what Capper had cobbled together from conversations with the vet, the landlord, the bailiff, barmen, and the police chief. By then, Capper had an account number and an account name he asked the banker to corroborate.

"I had enough pieces of paper with me to show that I hadn't just walked in off the street. The manager verified the Mastros had a current account, when and how it had been used, how frequently they had used their ATM cards, and the names of the signatories."

Throughout his investigation, one of the best sources Capper cultivated was a court-appointed translator. (Capper asked that I not name her, so I'll call her Laurel.) The police chief in Annecy once mentioned her offhandedly. Capper looked her up. They met and had a glass of red wine. He told her true stories about his thrilling life as a private detective. They discussed art. The Mastros hardly came up.

At a subsequent meeting, Laurel, unprompted, offered that Mastro was to be escorted by French police to a different branch of Société Générale to remove the contents of safe-deposit boxes he had leased. The law enforcement agents required Mastro's presence, because the boxes were in his name. Laurel knew it would be on Friday, but she did not know what time. So on Friday morning, Capper visited the bank and spoke with the manager.

"I asked him if we could speak privately about a complicated matter," Capper said. "That piqued his interest. I told him who I was and that I knew Mastro was due to come in later that day to access his safe-deposit box—again, offering instead of asking for information. I asked what time he expected Mastro to arrive. He said between two and four."

Capper hung around the bank, and when the Mastros appeared, they and their police escorts headed downstairs to the vault. Capper waited upstairs. A few minutes later, he heard a deep humming underground.

"Drilling," explained Capper; the police did not have the keys, so they had to break open the boxes. (No one knew where the keys were; perhaps they'd been confiscated.) Capper wanted to get documentary proof that the Mastros were in the bank that he could send to Rigby.

He and Hall, who had flown in from London, stepped out of the bank and scanned the area for a decent vantage point. There was a hair salon on the second floor of a building across the street.

"We walked in and explained in really bad French that we didn't need haircuts, but wanted to step out onto the balcony, because we were fascinated with the local architecture, and in particular the rooftops, and wanted to make some sketches," Capper said. It worked.

"There we were in a salon full of twentysomething French-women cutting hair, and they thought we were absolutely mad—two English guys drawing rooftops. They just sort of humored us." The hairdressers brought Capper and Hall espressos. They sketched. Soon, Mastro and his law enforcement handlers walked out of the bank, and the two men took the photographs they needed. (Simply documenting Mastro's movements was valuable to Rigby, because it could help provide clues to assets.)

Although they did not learn until later the contents of the safe-deposit boxes, it was spectacular investigative work. They were reporting in real time to their American client how the Mastros were being forced to disclose their hidden assets in France.

French law enforcement agents eventually seized everything they found—jewelry, haute couture, cash, the contents of bank accounts—and turned it over to the FBI, with whom Capper had developed a close relationship during his investigation.

Later, when he spoke with American agents about Mastro, he used it as an opportunity to collect more clues.

"We weren't horse trading, but I did learn a lot by asking questions. 'How much stuff have you found?' 'Have you found the rings?' 'What was in the safe-deposit boxes?' 'What will happen next?' Over the course of five hours, I got quite a bit of information out of them. I just had to remember it."

Among the assets that were cataloged in March 2013 by the bankruptcy trustee and appraised by the Gemological Institute of America were nearly three hundred pieces of jewelry and other goods worth more than $3 million: gold rings with South Sea–cultured pearls and onyx ribbons, Oyster Perpetual Cosmograph Daytona-style Rolex watches, a fourteen-karat yellow-gold diamond-encrusted bracelet with emerald cat eyes, Tahitian pearl pendants and earrings, diamond-encrusted gold rings in intertwined ribbon motifs, yellow-gold earrings lined with oval cabochon jadeite, Art Nouveau pendants, fourteen-karat gold monogrammed cuff links finished in a combination sandblast and pavé with fifteen diamonds in each link, watches by Patek Philippe, and white-gold tennis bracelets with princess-cut stones.

When the U.S. Marshals Service auctioned much of the jewelry in July 2014, the tally was only $350,165, about one-tenth its appraised value. The most popular item, which received 262 bids, was a fourteen-karat white-gold ring inlaid with eight marquise-cut emeralds and eight round diamonds set around a cabochon apple-green jade. It sold for $3,650.

Rigby, who ultimately recovered the proceeds from the auction from the marshals, was furious the marshals ever had possession of the jewels. "Where do they get off taking our property and selling it?" he asked.

This kind of jurisdictional turf battle is not uncommon in international cases involving bankruptcy courts and law enforcement officers who often fail to coordinate, Rigby told me.

Rigby separately sold Linda's diamond rings as well as Dolce & Gabbana trench coats; Roberto Cavalli, Christian Dior, and Gucci sunglasses; Cadolle furs; and rare coins.

Capper said that even though he did not personally transport a planeload of Cartier watches and Ferragamo handbags to Washington State, the information he dug up was invaluable.

"We were the front line of information. Rigby, who was

under pressure and spending the creditors' money, was on the phone with us and emailing us constantly. 'Where's our money?' 'What have you found today?'"

Capper continued, "Because considerable measures were taken before we were hired—seizures of Mastro's real estate and discovery attempts—Rigby's legal fees were piling up."

So when Capper fed Rigby details that journalists and police detectives did not have—bank account numbers and balances, photographs of Mastro walking out of a bank with police officers who had drilled into his safe-deposit box, descriptions of garbage bags full of unworn Dior dresses they watched being dragged out of Mastro's home, receipts from a pawnshop—he had answers to the creditors' anxious questions.

Mastro's subterfuge, his money-laundering machine, was hard assets: jewelry, cash, coats, and so on, making him hard to detect.

"You piece together all the flaky details—the gold sales and the vet bills and the rental cottages—and you start to get a narrative of what they did with their time, how they got money, how they spent money, how often they went to Geneva."

Rigby corroborated Capper's assessment. "Their investigation was magical," he said. Many of their leads came from what Rigby describes as a savvy ability to have smart conversations with witnesses, not by sweet-talking them, but by "asking questions to engage people, to draw them in—not expecting to get easy answers."

In addition, Rigby, who in thirty-five years had never worked with private investigators, was surprised by their youth: both men were under forty. "When I met Daniel and Robert in a hotel room near Annecy, I expected retired detectives from Scotland Yard," he admitted.

Near the end of the investigation, one of Capper's sources told him there was to be a hearing soon in a French criminal court during which Mastro would be interrogated about his assets.

On the day of the hearing, Capper posted himself outside the prison and waited. His plan was to follow the prison van to the court, sit in the gallery, and watch. But when he got to the prison, he learned the van had left three hours earlier for Lyon and that the session was to begin within the hour.

Capper sped to Lyon and rushed into the courthouse, only to learn the hearing was closed to the public. By then, he said, "a real posse of journalists had assembled and were fussing about. The little court clerk's office was surrounded by reporters firing off questions: 'Can I get a transcript?' 'Is the hearing adjourned?' Question, question, question. Soon the blinds came down. We were shut out."

Capper found a local café, had a sandwich, and returned two hours later. By luck, the chief clerk had gone to lunch, and her junior was on duty. Capper joked with her about the pandemonium surrounding the Americans, and at the end of the conversation she showed him a copy of the transcript, which he photographed with his phone. Among other details, it held information on what Mastro had stored in the safe-deposit boxes.

I asked Capper how he persuaded the clerk to show him such a confidential document. He answered with a cryptic analogy that left me as curious and mesmerized as I suspect the clerk was.

"If you're in a nightclub and there's a girl you like, you don't walk straight up to her and say, 'What's your number?' If you do, you shouldn't be surprised when she says, 'Go away. You're weird.'" Instead, he continued, "Hit on her friend. Pick your moment. Don't go in with the killer question. That can wait."

As of this writing, Rigby has collected about $24 million in assets formerly controlled by the Mastros, most of which has come from real estate sales, for his clients, the creditors.

Approximately $1 million of those assets was discovered in France and can be tied to Hall and Capper's work, Rigby told me. He wishes he had recovered more but understands that recovering everything would be impossible.

The Mastros, who never faced charges in the United States, still live in France off about $2,000 in monthly Social Security payments and donations from friends, according to news reports, but Rigby has his doubts.

"Mastro is brilliant, polite, and elegant," said Rigby, who deposed Mastro near the beginning of the bankruptcy proceedings. "He is smarter than his lawyers, and his thinking is not restricted by the law, which builds neat little boxes around concepts. But there are cracks and crevices in those boxes that Mastro finds."

He added, "Mastro is a master of misinformation." Indeed, said Rigby, some of his wealth was a facade. For instance, a lab in Los Angeles Rigby commissioned to appraise Linda's rings determined one was much less valuable than he assumed. "I thought we'd get a million for it," Rigby said. "We only got a few hundred thousand."

Mastro may still have millions of dollars in Swiss bank accounts. "To this day," Rigby conceded, "I don't know if he has a significant amount of money—or nothing."

*H*all and Capper's investigation is a prime example of the role private detectives play in ensuring that U.S. court systems function. In this case, the two investigators helped resolve claims by creditors seeking payments from people who argued, implausibly, that they were unable to meet their financial obligations and needed the courts to intervene.

The American bankruptcy market is sizable. Between 2005 and 2017, for instance, nearly 13 million consumer bankruptcy petitions were filed in the federal courts, according to the Administrative Office of the U.S. Courts. Of those 13 million, 8.7 million (nearly 70 percent) were filed under Chapter 7—the chapter under which, in part, Mastro's proceeding played out.

Of course, most debtors are not wealthy, sophisticated fugitives, and therefore funding (provided by creditors) is not always available for international manhunts.

And so investigators like Hall and Capper use their skills for other clients in other, sometimes related fields. Pre-transactional due diligence, for instance, requires many of the same detective tools as asset hunting. In that context, investigators help provide confidence to investors that their choices are, or should be, sound.

The global real estate investment market, in which Mastro operated for decades, is vast and can be murky, especially because high-end properties in places like New York and London are increasingly owned by opaque corporate entities rather than individuals.

In 2017, there was roughly $8.5 trillion in what is described as the "professionally managed" global real estate market, according to MSCI Inc., which analyzes and provides tools for various investment vehicles and sectors. The "most significant absolute change" in a single country, according to MSCI data, was in the United States, where the size of the market increased by $244 billion. (America is the largest real estate market in the world, followed by Japan, the United Kingdom, Germany, and China.)

In other words, there is a lot at stake, and investors are keen to make sure their partners are legitimate. To do this, they often spend resources on investigators not only to conduct due diligence before a deal but also to recover lost assets after a deal goes south—the way Rigby did with Mastro's creditors.

American federal prosecutors indicted Mastro on forty-three counts of bankruptcy fraud and money laundering, but the Mastros fought, and won, an extradition request and stayed in France. Rigby learned of Mastro's indictment when he read about it in the newspaper. A French court ruled against the request by the United States to extradite Mastro, citing humanitarian reasons: age and illness. Although the Mastros have considerably fewer resources thanks to Hall and Capper's work, they still live openly in France.

5.

BARE FEET

Julian Fisher was downing a Tusker beer by the hotel pool in Mombasa, Kenya, when the army chief called his mobile phone.

Fisher had been waiting two hours for the call. He was used to waiting. Much of a spy's life overseas is waiting—for meets with agents, operational engagements, instructions from station—especially on foreign assignments, where heightened surveillance does not permit much movement between hotels and rendezvous points.

For most of the past decade, Fisher had been a British intelligence officer. But today he was not spying. He was done with all that. He now runs a private intelligence firm based in London called Africa Integrity that helps corporate clients assess security risks and business opportunities in the Middle East and Africa.

As a consultant, he has discovered that a French telecom

company in Dakar was securing government contracts through bribes. He has advised American executives who fear they are targets of kidnap and random schemes on how to drive defensively in Mogadishu. And he has exposed a gang of pharmaceutical drug counterfeiters in Tunis. He is a man who knows how to deal with the powerful and influential across Africa.

After a few reluctant pleasantries, the chief invited Fisher to his office.

"Meet me at—" Fisher got an earful of wide Kenyan vowels.

"I beg your pardon, where . . . ?"

"Neeaftitti house," is what it sounded like. The chief hung up.

Fisher has spent years in Mombasa, Nairobi, and other parts of Kenya, and he understands the nuances of Kenyan English. He decided the chief said, "Near the statehouse."

There was an army command post near the statehouse, and he decided to head for that. He had one shot at meeting the army chief, who was known not to be inclined to answer his phone.

Fisher's client on this project was a European real estate developer, financier, and philanthropist with dozens of properties and grant recipients throughout East Africa—shopping centers, office towers, environmental and health care projects, schools. A rash of recent attacks and threatening anonymous phone calls had made the client anxious about his holdings in the region.

In 2011, there were terrorist attacks and kidnappings

orchestrated by al-Shabaab, a jihadist fundamentalist group in East Africa, at churches and markets in Kenya; victims included a British tourist. Kenya responded by sending troops to Somalia to battle fighters from al-Shabaab, which the U.S. State Department labeled a foreign terrorist group in 2008.

Al-Shabaab countered with new attacks in Kenya. In 2013, al-Shabaab took credit for killing sixty-seven people at the Westgate shopping mall in Nairobi. Their ranks hold jihadis who fought with al-Qaeda and bombed embassies in Dar es Salaam and Nairobi in 1998. Separatist movements like the Mombasa Republican Council, which seek autonomy for coastal areas, had also been accused of violence in Kenya.

Fisher, who in his former career trained nascent African intelligence agencies—"Lesson number one: 'You can't beat up everyone'"—was retained through his client's British counsel to identify targeted properties and those responsible for the threats and to see how serious they were. His fee was $25,000, half up front. ("It was not enough to mitigate the dangers and discomforts of this assignment," Fisher told me.)

As he put down his phone, Fisher looked up and noticed that he was alone at the hotel. Stories of kidnappings and murders had kept tourists from the Kenyan coast. The pool was still, most of the rooms empty. Waiters and housekeepers, who outnumbered guests, fidgeted in boredom. He paid his tab and went outside, where he caught a cab.

As his car made its way through town, not far from the two pairs of gigantic aluminum elephant tusks that arch over Moi

Avenue, one of Nairobi's main streets, the taxi suddenly swerved around a jaywalking pedestrian, a lifeless hammerhead shark slung over his shoulder.

Like every good spy, Fisher noticed but didn't react. Julian Fisher does not look or sound like the covert MI6 operative he once was. He is tall and gangly, studious and contemplative, and dresses in blue jeans, mock turtlenecks, and jogging sneakers. When I first met him a few days before in a pub in London, he unburdened himself before we'd ordered a second pint of lager. Talking openly about the adventurousness and absurdity of his life, it seemed, was cathartic.

He genuinely and regularly "tussles," as he puts it, with the moral dilemmas of investigative work—for instance, in balancing a respect for privacy protections with the necessity of finding information that benefits his commercial clients, who have vastly different motives (and use vastly different information-gathering tactics) than did his previous employer, the British government. I liked him immediately.

"What is information?" he asked as we drank our lager. "Is it a commodity? Is it free? What makes it public? What makes it private? Who owns it? Who decides what its value is? What makes it secret?"

For MI6, the U.K.'s foreign intelligence service, to classify information as top secret, Fisher said, it must be "obtained covertly and not recoverable by any other means."

For all his tussling, however, Fisher is an indispensable resource for companies seeking to do business or resolve disputes in East Africa. He has a keen understanding of the region's customs and languages. He has deep experience running agents and developing government and business contacts, and he has a true affection for life in the British Empire's former colonies.

As cross-border transactions—manufacturing, extracting, selling, distributing, assembling, outsourcing—increase, there is a growing need to evaluate and assess foreign opportunities. Companies seeking information on their prospective global partners can, of course, turn to government agencies. American companies wishing to vet foreign partners can ask the Commerce Department's Foreign Commercial Service and the State Department's Bureau of Economic and Business Affairs for help. Diplomats will visit warehouses to verify their existence or try to unravel the hazy structures of quasi-governmental corporate entities—all in the service of furthering American business interests abroad.

But there is also a massive private industry committed to this work that is carried out by investigators like Fisher, and also by law firms, former military and law enforcement officers, and accountants, among others.

Some private firms that do this work specialize in identifying links between their clients' foreign partners or subsidiaries and government officials. The U.S. Foreign Corrupt Practices Act and the U.K. Bribery Act seek to prevent American and British companies and their agents from bribing government officials or their relatives in return for business. Financial regulators sometimes call these government officials "politically exposed persons."

Fisher has little competition in East Africa. The neighborhood is corrupt, impoverished, and dangerous. Kenya ranks 145 (of 175) on Transparency International's Corruption Perceptions Index. In April 2015, al-Shabaab militants snuck over the Somali border and shot and killed nearly 150 students at Garissa University College in what has been called the worst terrorist attack in Kenya since the 1998 bombing of the American embassy in Nairobi.

*T*he taxi dropped off Fisher at a cluster of municipal buildings, and he eventually located the local army headquarters in a drab, concrete structure.

"Good afternoon," he greeted a man sitting behind a spare desk. "My name is Julian Fisher. I have an appointment."

The man directed him to a wooden bench in a room off the hallway. One hour passed. Then another hour. Then one more hour. The stench of open sewer filled the building. Flies buzzed in and out the windows; there were no panes in the sashes. The attendant had not looked up from his newspaper. The phone on his desk stayed silent. No other visitors had come or gone. A ceiling fan was motionless.

Although he was familiar with the rigmarole, Fisher grew frustrated. His client had arranged the meeting through an intermediary Fisher knew from his days as a spy to be reliable.

Finally, he approached the attendant, who seemed surprised he was still there. The attendant exchanged a few words with someone on the other side of a door. Fisher was ushered in and directed to a single folding chair facing the desk.

The army chief was wearing a full military uniform with a beret, remembers Fisher. "He shook my hand, sat down without looking at me, and stuck his naked feet out from under the desk."

In Islam, Fisher knew, it is disrespectful to show your bare feet. "He seemed to be signaling to me that he was a person of import and that I was not. In Kenyan parlance, he was the Big Man, and I was his supplicant."

It is rare for investigators to disclose the identities of their clients. In this case, however, the client told Fisher he could use his discretion.

Most Kenyans are Christian, but the country's Indian Ocean coast, where Mombasa sits, is home to most of the country's Muslim population. The majority of Kenyan Muslims are Sunni, but Fisher knew the army chief was Shia, and he would use these facts to establish his authority: in addition to his business interests, his client was a leader in the Shia community and revered by local leaders.

"I bring greetings as a direct representative of El Haji," said Fisher, who asked me not to disclose his client's real name. (El Haji is a title sometimes granted to a respected elder in the Muslim faith who has completed the pilgrimage to Mecca.)

The chief's face went blank. He recoiled his bare feet.

From then on, Fisher refrained from using the first person, and in every sentence he referenced his client: "El Haji would like to know . . ."; "It is El Haji's wish that . . ."; "Could you explain to El Haji why . . . ?"

At first, the chief tried bluster to avoid answering any of Fisher's questions. He then calmed down but spoke in vague terms without saying anything concrete. He provided commentary that was practically word on the street, telling Fisher stories that had been covered in the media or common gossip about the political situation in the region.

Fisher feigned boredom and waited for the chief to talk himself out. "I suspect El Haji knows all that from reading *Taifa Leo*," he said, referring to Kenya's largest Swahili paper.

The chief paused; he understood now. The power dynamic in the room shifted. Fisher, as his client's representative, was now calling the shots.

The chief began answering Fisher's questions deferentially. Fisher asked the chief to help him identify anyone who had a motive to attack his client's properties.

The chief swiveled in his chair and reached down to a large steel safe on the floor. He spun the dial, swung open the safe's door, removed a thin folder that contained a few sheets of paper, and handed the file to Fisher.

The chief was making an effort to impress him. As Fisher read through the pages, he realized that the chief was disclosing national intelligence secrets. At the top of one page were the names of backers of a local militant group, and Fisher recognized the names of three Kenyan politicians.

"It was an extraordinary moment," Fisher admitted. "I had been shown evidence that Kenyan officials were bankrolling a local militant group with Islamist connections. It was the kind of information that in my previous life I would have immediately disclosed to my employer. Exactly the kind of information, in fact, that I was paid to uncover."

At the same time, he wondered if he had an obligation to alert British intelligence. "If this type of information was being withheld from British and American intelligence services, which it probably was, it would impede all the liaison efforts to contain the terrorist threat in Kenya. But if I disclosed the links between the politicians and the militants, would I be defaming the politicians?" He had no corroboration, no proof. He contemplated his obligations.

"There is no two-way flow between intelligence services and their former officers," Fisher told me. "What gives a service the right to expect information to be given to them? Was this a matter of national security? Would sharing the information be a breach of trust between me and my client? How would I know that it would be handled discreetly? Experience in the service taught me that 'need to know' is not always strictly observed. What thanks would I get for sharing this? It might advance the career of an officer that I gave the information to, but not my status."

But Fisher was not meditating on ethics. Although he grapples as much with it in his current career as he did in his former, his thoughtfulness does not interfere with his pragmatism.

"El Haji will want to know how the militants and their political backers were identified" by the chief's intelligence officers, Fisher said, and what their relevance was.

The chief explained that when low-ranking Islamist operatives were arrested for minor terrorist attacks, a judge set bail for them, and "when the detainees were picked up, secret police followed the men who paid their bail home," said Fisher. In some cases, through this surveillance, the police were able to establish that the men who paid the detainees' bail worked for Kenyan politicians.

Without taking any notes, Fisher handed the folder back to the chief and thanked him.

"I have given you something," the chief said. "Now you should give something in return."

Fisher expected this. He had been here before. He could parry the bribe request in two ways. He could feign naïveté and force the chief to become increasingly explicit—asking for "a chai," "a dash," "a small soda," "to show gratitude," and then "to make a payment."

Or he could refuse on the grounds that El Haji had not approved such payments, which acknowledged the possibility of one, and offer to pass the request on.

Fisher chose the second option. The chief withdrew the request.

"El Haji thanks you for your time and effort," Fisher said.

The chief, conceding defeat, put on his boots and walked Fisher to the red-dirt courtyard, where his taxi was waiting. The chief shook Fisher's hand, then bear-hugged him goodbye.

"His attitude was a mix of disappointment, because he had claimed no bribe, and relief, because I was leaving."

When he returned to London, Fisher drafted a report to his

client drawing on what he had learned in Mombasa and through other interviews he conducted in Nairobi and elsewhere in Kenya, including the names of the politicians he'd seen in the chief's office. Before submitting his report to his client, however, Fisher deleted the names.

"I had no way of proving that I had seen the document or even that it existed," he said, reiterating his fear that he would be targeted for defamation if the report was ever leaked.

And he knew the army chief would not produce it again if Fisher's client challenged him to validate his reporting. Instead, in the report he made reference to "high-ranking local politicians active in one of the coalitions competing in the Kenyan election."

Fisher did not share that information with the British or American intelligence services. "Client loyalty won out over loyalty to my old service or to the old queen-and-country nonsense."

But the report was still valuable. "The significance, in my assessment, was that the activities of the militant group were driven by political expediency rather than ideology, and they likely were not even aware of my client's activities in the region," Fisher said. "While other groups may have represented a specific threat to El Haji, I concluded that this particular group did not and that its activities would likely cease after the election."

Fisher's theory proved correct. The local violence was not part of a larger movement but one-offs funded by local politicians who sought seats in certain neighborhoods where they could claim to be tough on crime in the face of danger.

"This case illustrates an important, if obvious, point: that raw intelligence is not much use to anyone. Its significance often lies in the analysis, which in turn raises questions about whether intelligence services should have analysis sections. The U.K. does not; the U.S. does," Fisher said.

He added, "For private intelligence consultancies," like Fisher's, "it's all about the ability to interpret raw material, and that requires cultural attunement and contextual knowledge. A nonspecialist intelligence operative is like a pair of scissors with only one blade."

El Haji reacted sanguinely to Fisher's report. "I don't imagine he was surprised to learn about the link between politics and violence in Kenya. The real value of my intelligence was the prominence of the names on the chief's list, national as well as local in standing. And even though I did not disclose those names to my client, I was able to be more confident in my analysis that the group was more a political tool than a genuine, specific threat."

Fisher's report also included what he calls "actionable points," suggestions for how his client should improve tactical intelligence gathering. His report carried the following headings: terrorism, political instability, criminality, corruption, and fraud.

"I suspect they took significant action on the third and fourth points, and while I assessed collateral threat from the first two, I concluded there was no imminent threat to El Haji's interests from them. Targeting methodology for groups like al-Shabaab evolves rapidly, so I recommended El Haji appoint an intelligence asset to be based in Nairobi to cover day-to-day developments."

When Fisher left the chief's office, he caught a taxi, which pulled into the road and settled into Mombasa traffic. Fisher expected to be placed under surveillance for the remainder of his stay.

He was—and he laughed when recounting the chief's efforts. At a restaurant in Mombasa during lunch the next day, the waiters were oddly attentive and lingered nearby. Two burly figures sipped sodas at the next table, while the other tables were empty. Later, Fisher was propositioned by a prostitute who knew El Haji's name.

This harassment, said Fisher, was likely designed to gain some advantage over Fisher and to demonstrate the chief's power. "It gave the chief a small victory in return for making him feel uncomfortable," he said. "It was designed to intimidate me rather than gather information on me."

In reflecting on this saga, Fisher remembered that as he was approaching his hotel in the taxi after his meeting with the army chief, he considered the absurdity, and precariousness, of the ordeal.

He noticed the rear window of a brightly painted *matatu*, a commuter minivan. Next to a yellow sticker that read "God Is Good All the Time" was an advertising slogan finger-painted in a film of dirt: "We buy old guns."

Fisher told me that his client, who was no stranger to East Africa, was impressed with his work, especially his conclusion that the client's properties were not targeted because of who owned them. Fisher's intelligence persuaded El Haji not to divest from a region that has trouble attracting foreign money.

Fisher was valuable to El Haji in this case, in part, because he has deep networks he had developed as an intelligence officer that he brought with him to the private sector. Investigative work outside the United States is more difficult because there is less information that is available to the public.

Private eyes abroad often depend on discreetly speaking with trusted sources, not database research. Even when government websites are available in the developing world, they can be undependable.

Fisher was aware of this when he once helped an English pharmaceutical company investigate whether it was in violation of U.S. and UN sanctions that had been imposed on doing business with Sudan because of atrocities carried out by government forces and proxies. The client wanted Fisher to investigate its own supply chain and learn how its distributor in Sudan got a Sudanese business license.

His first step was to establish that a Sudanese corporate entity was legitimate. There is no web-accessible database for this, so Fisher sent a trusted investigator of Sudanese descent to Khartoum, Sudan's capital, to obtain copies of the company's registration records, which, he hoped, would include its name, its date of formation, and a list of directors and shareholders.

Fisher knew this was a delicate exercise, because it was likely that any inquiries within government agencies in Khartoum would make their way back either to the people who ran the distributor or to suspicious government officials, including intelligence operatives who monitored requests for corporate records like these.

In most African countries, as in many developing countries around the world, explained Fisher, this kind of inquiry requires a healthy dose of paranoia: you must assume you are being watched.

Fisher's contact was able to get a copy of the corporate record from a trusted contact, ensuring that the inquiry would not leak. The record showed the company was legitimate, which gave the pharmaceutical company comfort it was not circumventing international sanctions.

*C*ircumventing conventions was of no concern to two employees of a real estate development firm in Dubai whose white-collar smash and grab was stymied by a journalist turned private eye from New York.

6.

SNOW IN THE DESERT

The sheikh was nearly out of ideas. His resources, his title, his connections, his tribal roots, his nobility, his education, his stature, his wealth—none had been helpful. Two employees of his real estate development firm had stolen millions of dollars from him, and although a European law firm had unraveled the scheme and revealed who the culprits were, it could not find them. They had fled; their apartment abandoned; their possessions sold or shipped away.

This was embezzlement, plain and simple, a subset of old-fashioned theft. There had been no force—not even digital force, such as hacking—no threats, no identity theft. This was not gangsterism or an elaborate confidence game. There was not even any deception. It was more like criminal opportunism.

Employees with inside access to customer accounts had diverted company assets to their own accounts. The assets were from Dubai, the largest city in the United Arab Emirates, which had become a global hub for routing capital, labor, materials, and contraband.

Modern Dubai, on the southeast coast of the Persian Gulf, was built on boosterism and speculation, and it had thrived. Thirty years ago, the emirs, understanding that their oil reserves were finite, made a concerted effort to attract financiers and developers to their city. Prospectors started to buy condominiums in towers that had not yet been built on ground that had not yet been broken.

Since then, Dubai has become a sprawling metropolis of rebar and concrete and glass and shipping containers, built on sand and landfill, soldered together by foreign laborers and indentured servants. The country has become a symbol of waste and exuberance; the Burj Khalifa, at 828 meters, is the world's tallest skyscraper. (Only two generations earlier, Dubai's inhabitants, many of them nomadic Berbers, had lived in homes made of palm fronds.)

But in the middle and late years of the first decade of the twenty-first century, global capital stumbled. Many borrowers defaulted, and many of the city's cranes froze. The securitization schemes were unmasked. Broken promises and unrealized dreams shimmied into the desert. Dubai was not immune to the global bear market.

It was in the wake of this downturn that the sheikh was robbed. His fortune—built not only on real estate but also on the exploitation of natural resources—while chipped, was not shattered. He knew he would recover financially. It was his ego that was likely damaged most. But he was determined to track down the suspects: Bijan, who helped run his firm, and Maya, Bijan's underling. (There was no romance, only larceny, and those are not their real names.)

A few months before the sheikh discovered this fraud, a thirty-two-year-old journalist in New York City was growing restless. Eric Latimer, a reporter for a daily newspaper, was enterprising, outgoing, and versatile. He wrote about political corruption in Brooklyn, reported on counterterrorism tactics used by the New York City Police Department, and covered the city's criminal courts. His byline was prolific. But he sought greater fulfillment.

"I wrote about investigations, but I did not do investigations," said Latimer, who has bright, piercing eyes and an easy smile. "My job was to find out what other people were doing, and I hated that."

He also understood the media was likewise going through a rite of passage, and investigative desks at magazines and newspapers were being downsized and defunded. Seeking some old-fashioned muckraking—Jimmy Breslin was an inspiration—Latimer set out to become a private eye. A friend introduced him to a former military intelligence officer who ran a small detective firm based in the Midwest, and within weeks Latimer had arranged a job with him working from New York.

It was not only his journalism credentials that made him an attractive candidate for investigative work. After college,

Latimer traveled widely, often alone and for months at a time. He worked as a commercial fishing deckhand off the coast of New Jersey and traversed the Amazon, where he slept in a hammock on a riverboat. He surfed in Peru and hiked the Andes.

"I learned to operate in places that would make a typical American nervous," he said. Latimer's new employer sensed this and within a few weeks called him with his first big assignment. An internal investigation by a British law firm had concluded that Bijan and Maya had stolen funds from the sheikh's company.

The client would not disclose much more. But Latimer didn't need many details, because he had not been tasked with uncovering any wrongdoing, which is more commonly why investigators are engaged. Instead, he was simply asked to find the two people who had perpetrated the crime.

With the help of an investigator in Dubai, the sheikh had already assembled dossiers on the two suspects. Bijan was Iranian—the U.A.E. has one of the largest Iranian diaspora communities—and had lived for many years in Dubai, where he had been a respected member of the business community.

The local investigator in Dubai had tried to find Bijan and had been unsuccessful, so the sheikh decided to prioritize Maya, a British citizen on whom they had a solid lead. While in Dubai, she had dated a man named Talal, who worked as an instructor at Ski Dubai, the largest indoor slope in the world.

Ski Dubai is its own ecosystem, a frozen structure in the sweltering heat. It is jerry-rigged onto the Mall of the Emirates, which holds more than six hundred retail shops. Rising two hundred feet, Ski Dubai's six slopes are carved into six thousand tons of artificial snow made from desalinated seawater. At a cost of more than $400 million to construct, it is a fantasy of enormous proportions.

The local investigator had called Ski Dubai and interviewed some of Maya's former colleagues. He developed two leads during these conversations: Talal was Lebanese and had recently left Dubai. (The Lebanese played a significant role in the development of modern Dubai. To this day, Lebanese who leave their country often migrate to the Gulf states.)

The investigator and the sheikh had developed a hypothesis to find the kind of bolt-hole every investigation needs: Talal had returned to Lebanon and brought Maya with him, and they lived near one of the few ski slopes on Mount Lebanon, the mountain range that spines the country.

To test this theory, the sheikh had a Lebanese contact ask discreet questions of some locals, including shop owners, merchants, investors, community activists, and lawyers.

Talal, he'd learned, had been spotted in a small town in the mountains northeast of Beirut, and he was living with an unidentified blond woman who they suspected was Maya.

The sheikh's lawyers had made some progress in the case—uncovering emails and bank transactions incriminating Bijan and Maya—but they could not find where the thieves had gone. They determined that a civil lawsuit would be useless. Even if the sheikh won a judgment, collecting the debt would have been unlikely. It would have been useless to report the theft to regulators in the U.A.E. (of which Dubai is one emirate among seven within the country), who are often slow to act, uninterested, or unable to pursue cases that fall below a certain monetary threshold. Law enforcement officers within Dubai had not been able to track the fugitives, and even if they had, the authorities had no jurisdiction outside Dubai, Latimer told me.

But none of that mattered to the sheikh, Latimer said. His goal was not to be made whole, or to serve Bijan and Maya with a complaint or a subpoena, or to threaten them. He had another idea, which involved Latimer.

Latimer flew to Dubai and spent a few months preparing for his mission: to find Maya in Lebanon.

He met with the sheikh in his office in a tower high above the city. The sheikh, who wore a crisp white robe and keffiyeh, offered some Scandinavian tobacco to Latimer. They dipped together like a couple of lacrosse players at a fraternity party.

Latimer also met with the local Dubai investigator at a Syrian hookah bar near the waterfront. They smoked *shisha* (flavored tobacco) and planned Latimer's trip to Lebanon. They also conducted surveillance in Dubai with some former British Special Forces who had more experience tracking fugitives. They were looking for members of Bijan's family who might lead them to Bijan.

During one stakeout, Latimer thought he saw Bijan enter an apartment building. He called the Dubai secret police, who did not storm the apartment, as Latimer, still a newcomer to the field, figured they would. Instead, they walked up to the apartment and knocked on the door. Bijan was not there. It was his brother, who claimed to know nothing of Bijan's whereabouts.

Before leaving for Lebanon, Latimer snorkeled in the Gulf of Oman off a beach frequented by Russian tourists whom Latimer suspected were not simply on holiday but, he speculated, engaged in shady dealings of their own. Latimer caught and filleted a *hammour* (a kind of cod) on a fishing boat near Abu Dhabi and visited the observation deck of the Burj Khalifa.

*A*t Rafic Hariri International Airport in Beirut, Latimer hailed a taxi, and as he was driven to his hotel, he observed the lingering remnants of war, of a city traumatized by two generations of strife—collapsed buildings, bomb sites, facades riddled with bullet holes, abandoned lots.

The next morning, Latimer met with a local fixer named Rafiq, who would help him navigate the terrain of the Mount Lebanon range, translate, and drive. Rafiq was heavyset, in his late twenties, and rugged but friendly, Latimer told me. He described Rafiq as a "normal Lebanese kid." When he arrived at Latimer's hotel, he was sneezing and wheezing: Rafiq had a terrible cold.

Fixing was not Rafiq's full-time job (he was a bouncer at a nightclub), but he needed the money, spoke perfect English, was from the Christian north—where Latimer was headed—and had some law enforcement connections.

Although Latimer had been to Dubai and Turkey, he had

no other experience in countries that would fit into even the broadest definition of the Middle East. He needed Rafiq.

They climbed into Rafiq's silver Peugeot sedan and drove forty minutes north along the Mediterranean coast to Byblos, Rafiq's hometown. Byblos is often called the oldest continuously inhabited city in the world, and the origin of the English alphabet was discovered there. Among the oldest Phoenician cities, Byblos is designated a UNESCO World Heritage Site and includes Bronze Age temples, Persian fortifications, a Roman road, Byzantine churches, a Crusades-era citadel, and a medieval and Ottoman town. Built on a cliff of sandstone fanning around an ancient port, Byblos is dotted with crumbling stone walls, lone columns, and amphitheaters.

Although its Shia population is expanding, most of Byblos's residents, who are fond of referring to themselves as Phoenicians, are Maronite Catholics, who make up about 20 percent of the Lebanese population.

Latimer found a local hotel and used the city as a staging ground for forays up into Mount Lebanon, which separates the Mediterranean lowlands to the west and the Beqáa Valley to the east. Latimer was there during the summer, when sun scorches the coastal landscape, but in cold months snow blankets the mountaintops. (The horizontal white band on the country's flag is meant, in part, to symbolize this.)

The next morning, Latimer and Rafiq drove up narrow, unmarked roads through the countryside covered by small livestock farms and huts and through thick forests of pine and oak trees. Latimer was struck by how close together everything was. Lebanon is the smallest country in continental Asia—about one-third the size of Maryland—and about as many people (around six million).

Ten minutes outside Byblos, Rafiq pulled off the road into a small village to speak with a friend who, Rafiq suspected, would know of Talal. Behind the friend's home was a medieval stone barn, and standing outside were the friend and the friend's uncle, an elderly, bearded man who was cradling a large rooster under one arm.

While Rafiq spoke with them in Arabic, Latimer noticed a cockfighting pit nearby. He felt far away from home.

Rafiq turned to Latimer. He had good news. Talal, Maya's boyfriend, lived about ten minutes up the mountain, where he owned a large farm at the far end of town. His business was importing and selling foreign cars, not always legally.

Although Lebanon has a free-market economy, the country is stifled by bureaucratic waste, corruption, confusing customs procedures, high taxes, and archaic legislation, among other roadblocks to growth.

Latimer and Rafiq thanked the men and continued on their quest. Soon they arrived in a small, nondescript town and found a convenience store where Rafiq casually asked if anyone had seen a blond European woman. Yes, the proprietor told him. She lives a few blocks away and drives a large yellow car.

Struck by their good fortune, Latimer and Rafiq drove around town and within minutes saw at a crossroads not far from the convenience store, in front of a boxy, modern stone house, a yellow Hummer.

Modeled after military-issue Humvees (High Mobility Multipurpose Wheeled Vehicle), Hummers are broad, chunky vehicles that dwarf SUVs. Latimer wondered how a Humvee could navigate the narrow mountain passes; it seemed an incongruous and inefficient—not to mention conspicuous—vehicle in this isolated village.

Latimer decided not to approach the house immediately, but instead to familiarize himself with the area. They drove around town and then headed to the far end of town to find Talal's farm. Near the top of a ridge, they pulled over and asked a woman who was tending to a roadside stand if she knew of a farm that was owned by a Lebanese man who would have been seen with a white woman.

"It's right up there," she said, pointing.

Most investigations require a staggering series of failures before even a vague hint at a lead seems to emerge. Even though this was his first big case, Latimer was struck by how smoothly everything was going. It was hard to believe that they were on the right track, that one piece of evidence after the next was turning up so easily.

As they drove in the direction the woman had pointed out to them, Rafiq had difficulty maneuvering his Peugeot on the thin, unpaved roads. Near what felt like the end of the road, he turned off the engine, and he and Latimer got out of the car.

"I remember feeling like we were at the summit of Mount Lebanon," Latimer said.

They trudged a few meters uphill and approached a wooden fence. They looked down from the rim of a valley nestled among peaks and marveled at what they saw: on a field the size of several soccer fields, dozens of doghouses were scattered, and pacing throughout were German shepherds. Maya and Talal were breeding dogs.

For the first time, Latimer felt apprehensive, if not exactly fearful, about his mission: a yellow Hummer, smuggled cars, hundreds of German shepherds—all in a secluded hamlet tucked up a narrow road near the peak of Mount Lebanon. It was beginning to feel like a more mysterious mission than they'd originally thought. Latimer wondered if there would be more surprises.

He and Rafiq gaped for a few minutes. There was no one on the property. Only dogs. Latimer took out his camera and took photographs. They decided against trespassing and headed back to Maya's house. The sun was setting. When they arrived, the Hummer was gone.

They returned the following morning; still, the car was not there. Because there was nowhere inconspicuous to watch the house, Latimer and Rafiq found a small restaurant, sat down, talked sports, and smoked hookahs.

The next morning when they returned to the village, the Hummer again was nowhere to be found. Latimer did not want to knock on Maya's door without evidence that she was home. He returned to Byblos, toured the ruins, and perused shops selling marine fossils, for which the town is known. That evening, he and Rafiq had dinner and drove back up Mount Lebanon.

Finally, the Hummer was back. Rafiq parked a block away,

and Latimer gathered a manila folder of documents he had brought with him and considered his approach. His assignment was to take an arsenal of records—printouts of emails, spreadsheets, bank transactions, and financials showing Maya as a beneficiary (and a crook)—and personally hand them to her.

Also included in the packet was a note from the sheikh: "Please come back to Dubai and face the charges that are against you."

The purpose of the documents was to inform her that she was a thief, that the sheikh was aware of her crimes, that she could not hide, and that she was wanted in the United Arab Emirates.

This was not about vengeance. It was an exhibition of power.

At this point Latimer was afraid. "I've knocked on doors in East New York in Brooklyn to talk to witnesses," he said. "But this was different."

Rafiq waited in the car while Latimer walked up to the house and knocked on the door. No answer. Latimer waited a minute, then knocked again. No answer. He and Rafiq retreated, found a bar, and waited. When they returned, they knocked again, and again no one was home.

An hour later, they went back, and as Latimer prepared to get out of the car, Rafiq—now impatient, exhausted, still sniffling from his cold—intervened.

It was nearly dark and Rafiq noticed a child, perhaps ten years old, walking toward them. Rafiq handed the boy a few Lebanese pound notes and said, "Hop over that fence and tell the people inside we want to speak with them."

The boy put the bills in his pocket and climbed the fence on the corner, which wrapped around a yard behind the building. Latimer and Rafiq sat in silence. A few minutes later, the boy returned.

"They're coming," he announced with nonchalance, and walked away. Latimer approached the door apprehensively. A blond woman opened it.

"These are for you," he said without a smile. He did not introduce himself. He did not explain the purpose of his visit. He did not confirm that this was Maya. "My phone number is in there if you'd like to talk."

Later, Latimer admitted that he hadn't known for certain that this was Maya.

"My client and I did not really discuss how I would approach her," he explained. "I wasn't supposed to engage Maya or interview her."

Nor was he to tell her that he was delivering the papers on behalf of the sheikh. The intrigue was intentional, though Latimer acknowledged he should have had a plan in place that would have allowed him to confirm her identity.

The woman, expressionless, took the folder. She closed the door. Latimer turned and walked away. The interaction had lasted ten seconds.

Moments later, as he and Rafiq were driving back down the mountain, Latimer's phone rang.

"Why are you coming to my home?" yelled the woman frantically. "How dare you come to my home and bring me this! Who do you think you are?"

"If you can't calm down," Latimer replied, "I can't talk to you."

She kept screaming. He hung up. This was a triumph and a relief for Latimer. The call was confirmation that it had indeed been Maya who opened the door and taken the package. Latimer had accomplished his mission.

Five minutes later, his phone rang again. This time it was Maya's boyfriend, Talal. He was calmer. He requested a meeting. Latimer told Talal he would call him back. In a brief phone conversation with his boss in the United States, they decided against such a meeting, which could have been confrontational. Latimer feared Talal would be violent.

Latimer did not return Talal's call. He removed the Lebanese SIM card from his burner phone, and he and Rafiq returned to Byblos. Fearing Talal would track him, Latimer packed his bags, and Rafiq drove him to Beirut. The next morning, Latimer was on a flight to New York.

This experience launched Latimer's career. He now works for a global consulting firm where he is responsible for managing corporate investigations and compliance projects around the world that are staffed by dozens of investigators, which include linguists and accountants.

He still wonders what happened to the sheikh and the couple in Lebanon.

"I was never given the full context," Latimer said. "It was always a little strange; there is this part of me that still isn't sure who the bad guy was."

The fact that Latimer was carrying out a vendetta made both sides seem capable of nefarious behavior.

"The sheikh wanted the people who stole from him to know he knew they stole from him," explained Latimer. "And he spent real money to prove that point."

Latimer's Emirati-Lebanese adventure was a stark lesson in how business is transacted (and wielded, almost as a weapon) by people at the top of the wealth pyramid and illustrative of some of the egos and motives at play.

To be sure, this was an unorthodox assignment. Like all consultants, private investigators do not always choose their clients. Sometimes our work benefits the social good; sometimes we are the instruments of moral outrage.

I was once hired to interview dozens of people for one reason: so that the witnesses I called would tell their boss that a private eye had been asking tough questions about his reputation. I was, in essence, used as a tool for intimidation. As in Latimer's case, my goal was not to recover money or to find facts but to exert pressure in order to persuade someone to settle a lawsuit.

The sheikh did not use Latimer to achieve justice. He simply wanted the last word: "I know where you are, and I know what you did."

As far as I know, neither Maya nor Talal ever returned to Dubai.

7.

A LITTLE HONEST SHOOTING

Michael Gibbs circled the block in a rented Honda Accord looking for an inconspicuous place to watch Jim Spurn. In satellite images of the leafy, manicured neighborhood in a suburb of Dallas, there was a shoulder on the curve between Spurn's four-thousand-square-foot home and a playground at the corner. But Gibbs realized the images were dated. Most Google Earth data is between one and three years old. A sidewalk now lined the street. The only parking was in private driveways.

Gibbs drove around the block, then pulled into the lot of a Starbucks half a mile away. He leashed Oscar, a terrier panting in the back seat, and exchanged a few words with two of his colleagues, a couple sitting in a Ford Focus.

He and Oscar strolled toward Spurn's five-bedroom house, which sat on two acres fronted by clusters of azaleas, sage bushes, and agave plants. Gibbs noticed a Ford F-150 pickup parked in the driveway. He pulled out his iPhone, took a photograph of its license plate, and texted it to Susan Foley, a colleague of Gibbs's who was at her desk in an office building in midtown Manhattan. He then took photographs of a few neighboring homes. He pretended to have a conversation on his phone, gesturing as he normally would.

"Conspicuous behavior is inconspicuous," Gibbs told me later, with no hint of irony or sarcasm. As he walked and talked, Gibbs scanned for evidence of inactivity at Spurn's home: drawn window shades, spiderwebs or dirt or dust on walkways, switched-off lights, dried-out or overgrown grass, untrimmed hedges. From a distance (he did not walk onto the property), he inspected the truck for dirt and mud on the wheel wells.

Gibbs noticed a cluster of wrapped newspapers at the end of the gravel drive. The mailbox was open and stuffed full. He wanted to grab a letter or two, but because the mailbox was on the lawn, not on the public side of the property line, if he had taken one, even to look at, he would be trespassing and stealing private property.

Oscar barked. A woman in a Southern Methodist University cap jogged by and smiled.

Gibbs tried to determine whether Spurn's trash cans were placed on public property. If they were, he could collect Spurn's garbage. That was when he noticed a promising sign: a neighbor's recycling bin was sitting at the end of the drive, halfway into the street. Gibbs called Dallas sanitation services and learned that Thursday was trash day for this neighborhood. He would return two days later, in the middle of the night, before the garbagemen arrived.

Gibbs was also hoping to catch Spurn leave home (he knew what he looked like, because Gibbs had a photograph of Spurn from Facebook). If he did, Gibbs would text the couple in the Ford Focus, who would follow Spurn.

Surveillance—from the French *sur* (over) and *veiller* (to watch) and coined in the nineteenth century after the Latin *vigilare* (to keep watch)—has, in recent years, taken on new digital dimensions, of course, but law-abiding private eyes in the United States usually still do it in cars or on foot with their eyes and ears.

Gibbs had decided to knock on Spurn's door. He had memorized some information in anticipation of a conversation with Spurn, or a relative, a nanny, a contractor, a babysitter. For instance, he knew the house was built in 2002 on an abandoned golf course and that Spurn purchased it the following year for $1.2 million with a $950,000 mortgage from Wells Fargo. He knew the name of the local public schools, where the closest Walmart was, the average home value, even a few of the neighbors' names.

What Gibbs could not do, what he was not permitted to do, was lie. He could not impersonate. He could not deceive. State laws vary, but private investigators are generally not allowed to fabricate their identities or make up why they are interacting with someone. For example, if Spurn's nephew answered the door, Gibbs could say he was curious about local home values, but he could not say he was a real estate agent trolling for clients. These limitations rankled Gibbs, who had spent most of his life in law enforcement, where flouting laws in undercover operations is legal. Now he was in the private sector, and much of the tradecraft he employed as a federal agent was off-limits.

Gibbs tugged on Oscar's leash and made his way up the driveway. The jogger returned from her run. She slowed to a walk, removed her earphones, and approached him.

Things were about to get personal.

About once a month in my job, someone asks me to break the law. The request usually comes from a client who's desperate to resolve a dispute—a patent was infringed upon, or data was stolen, or a contract was breached, or employees were poached.

Sometimes the request comes from a company curious about its competitors' secrets: the location of a manufacturer, a supply chain process, a pricing structure. Sometimes the demand comes from a spurned spouse.

An attorney once asked me to bribe bank officers for account numbers and balances. According to New York State Penal Law, this might be charged as commercial bribery, unlawful possession of personal identification information, and larceny. Felonies.

A husband once asked me to hack into his wife's Facebook and email accounts and bug her phone, laptop, and car: computer tampering, computer trespass, and/or unauthorized use of a computer, and eavesdropping. Felonies.

A union once asked me to sneak into the manufacturing

plant of a company whose workers it was trying to organize: trespassing and burglary. Felonies.

I've been asked to track down flight manifests, figure out how to remove someone from a government watch list, steal trade secrets, and impersonate government employees. All felonies. All of which I declined to do.

Clients sometimes feel entitled to use unlawful tactics to right a perceived wrong. Some make such requests unwittingly, ignorant of the law. What makes these requests so shocking is that private detectives are usually hired by lawyers, who act as middlemen for their clients. As any intelligent lawyer knows, information collected illegally, even improperly, would not be admissible in court and could lead to civil or criminal charges or regulatory censures against the investigator and the lawyer. The client takes risks too. Asking a private eye to risk her freedom is a fool's errand. Still, it happens all the time.

*G*ibbs and I were investigating Spurn on behalf of a lawyer named Gabe Banning. I had worked for Banning twice, but I had never met him. His internet presence revealed little. A photograph on his firm's website showed him posing stiffly in front of a bookshelf of legal texts, and his biography highlighted campy accolades from *Super Lawyers*. But I got the sense that Banning was combative, militaristic in his adherence to hierarchy, and, I suspect, foolhardy. He ran the international arbitration and dispute resolution practice of a major

Los Angeles law firm, which means he was paid about $2 million per year to scuffle hard.

To begin the case, Banning and I spoke on the phone. Joining the call was Gibbs, who is youthful and polished, a former agent for the Federal Bureau of Investigation, and Foley, the researcher in Manhattan, who is a former librarian, sixtyish, introverted.

Banning was impressed with Gibbs. They had both been crime fighters, the kind of middle-aged white men who enjoy boozy lunches and conversations that aren't limited by political correctness. Gibbs was confident and raffish. He had spent five years on the Joint Terrorism Task Force, a partnership of federal and state law enforcement agencies. Banning had been an assistant district attorney in Queens. During the call, they gossiped about judges, a common topic of conversation among former prosecutors and law enforcement officers. Gibbs approached investigative work as a game and had the full-throated tone of a military veteran. He had worked in the private sector for more than five years since leaving the FBI but still rhapsodized about hunting dangerous criminals. Gibbs brought a prosecutorial bent to a job that doesn't always require that. He thought of Spurn as a suspect to arrest and indict, not the party in a civil lawsuit, which he was.

Banning prattled on for nearly twenty minutes, explaining the background of the case, which involved a constellation of American and British investors, Dutch coders, and Gibraltarian shell companies. Then he dazzled us with his legal strategy.

We were to unravel corporate relationships, help him establish legal jurisdiction in the United States, where courts are more sympathetic to international creditors, and track down assets.

Our subject was Spurn and his company, TopApp321, which Banning's client, 3-Draw, a developer of 3-D printers, had hired to build a mobile app and consult on its broader internet strategy. TopApp321 never delivered, so Banning sued for breach of contract and won a judgment for $9 million, which was, in addition to some fees, owed him. But Banning suspected TopApp321 would never pay its debt. And because courts have little power to enforce a civil judgment, that job was left to 3-Draw, its lawyers, and us.

Banning tried to negotiate a payment from Spurn, but Spurn and his lawyers claimed TopApp321 was insolvent and he had no other source of funds. Banning didn't believe him. That is why he called me.

If we could find TopApp321's assets, 3-Draw could grab them to satisfy the debt. There are many ways to do this. Although civil courts don't chase debtors, creditors are given the authority to track down money they are owed. We could repossess sailboats, tow Teslas, garnish wages, and drain Fidelity accounts. This is standard fare for corporate investigators.

Although it is difficult work, it is also among the most exciting, especially when the people we chase try to hide their wealth. And they do this creatively by forming trusts, moving funds offshore, hiding behind LLCs, and transferring real estate to spouses' and children's names, for example.

But Banning was not satisfied with finding houses and

sailboats and cars. "Send someone in posing as an investor," he told me. "Spurn is always working on some new idea. Get cozy with him. Eventually, you'll do due diligence and be given copies of his financials. Then we'll know how he values his businesses, the names of his partners, and what collateralizes his deals."

This was not the first time I was asked to assume a false identity. (Identity theft, false impersonation, criminal impersonation, unlawful possession of personal information. Felonies.)

My profession would be vastly different if I was legally allowed to portray the characters that clients have tried to cast me to play: an allergist, a heroin trafficker, a real estate developer, a truck driver, a rabbinical student, a pyramid scheme investor, a slumlord, the other man.

Before I could respond, Banning said, "And get me his bank accounts!" Felony.

Finding many assets—yachts, SUVs, summer bungalows, corporate entities, patents, wine, stock—can be easy. But the existence and contents of bank accounts are not public record. However, sometimes we get lucky by finding information about bank accounts in other records or by asking the right people the right questions. And therein lay our dilemma. I was determined to crack Banning's case, and I had two strategies: luck and crime. Let's begin with the crime.

*P*inkerton's National Detective Agency was formed near Chicago in the early 1850s to supplement the paltry resources of law enforcement agencies, which were in their infancy. At the time, crime fighting was routinely handled by the private sector, not any kind of police force. Allan Pinkerton, a Scottish abolitionist who provided security to Abraham Lincoln before the president's inauguration, professionalized an undisciplined industry, and he insisted that his detectives behave ethically.

"The profession of Detective is a high and honorable calling," according to a pamphlet published in 1867. A Pinkerton man "is an officer of justice and must himself be pure and above reproach."

The agency would take only cases that were "strictly legitimate and right, and then only for the purpose of furthering the ends of justice and bringing Criminals to punishment." Pinkerton called his men "operatives" and himself a "principal."

In the lawlessness that followed the Civil War, Pinkerton agents won assignments to track down and detain bank and train robbers, and they secured government contracts. Despite its initial success, in the following few decades, the Pinkerton agency was suffering, and the company lowered its standards in search of income. In one of his marketing pamphlets, called *General Principles and Rules*, Pinkerton provided justification for questionable behavior by the agency's detectives: "It frequently becomes necessary for the Detective, when brought in contact with Criminals, to pretend to be a Criminal . . . and to resort to deception."

But deception was among the most benign of his operatives' tactics. During the 1880s and 1890s, his undercover agents infiltrated and destroyed miners' unions, and they were often deployed as private security forces hired by industrialists. Their union-busting activities included violence and murder during a strike at Andrew Carnegie's Homestead Mill in 1892 in Pennsylvania.

The tragedy inspired lawmakers to ban the use of hired guards during labor disputes. (To this day, certain state laws regulating private investigators explicitly prevent some of these tactics.)

Among the most notorious of Pinkerton's operatives was James McParland. According to his biographer, Beau Riffen-burgh, whose *Pinkerton's Great Detective* unpacks some of the mythology surrounding McParland and the Pinkertons, Mc-Parland was a "master of evasion, obfuscation, and, at times, outright deceit," a man who "delighted" in "evading questions." He was "labeled a perjurer, an agent provocateur, and even a murderer."

Two generations later, the private detective became a central figure in American literature. Writers like Carroll John Daly, one of the first prose stylists to make use of the hard-boiled vernacular (tough, cold, urban), Dashiell Hammett, James Cain, and Raymond Chandler modeled their brash protagonists on nineteenth-century outlaws and cowboys.

They were wisecracking risk takers whose actions today would be indictable. In Daly's 1923 story "Knights of the Open Palm," the protagonist confesses, "I'm what you might call the middleman—just a halfway house between the dicks and the crooks. Oh, there ain't no doubt that both the cops and the crooks take me for a gun, but I ain't—not rightly speaking. I do a little honest shooting once in a while—just in the way of business. But my conscience is clear; I never bumped off a guy what didn't need it. And I can put it over the crooks every time—why, I know more about crooks than what they know about themselves. Yep, Race Williams, Private Investigator, that's me."

Nearly a century later, fictional private detectives in pop culture are no more principled (if equally entertaining)—even corporate dicks who spend more time in boardrooms than back alleys.

In *Michael Clayton*, the 2007 film starring George Clooney, a private eye who trespasses and taps phones is also an assassin.

In the television show *The Good Wife*, about a law firm, the in-house investigator, Kalinda Sharma, sleeps with police officers and steals their files, hacks into email accounts, breaks into office buildings, and impersonates witnesses.

(The show's writers hew close to literary tradition, giving Kalinda a malleable identity, a fondness for drink, and omnivorous bisexuality: Sherlock Holmes is a drug abuser whose sexual preferences are a mystery.)

Susan Foley received the photograph of the pickup truck in Spurn's driveway from Gibbs and ran the license plate number through the Texas Department of Motor Vehicles database. Licensed private eyes can usually obtain driving records on people they investigate only with a permissible purpose as defined by the Driver's Privacy Protection Act, a federal statute passed in 1994 that regulates and restricts access to DMV records.

The bill that became law was introduced after a woman was murdered in 1989 by a stalker who tracked her down using driving records and after opponents of abortion rights used DMV records to identify abortion patients and providers, many of whom were threatened, harassed, injured, and killed. Among the victims was Dr. Susan Wicklund, who wore disguises to avoid detection and whose daughter was followed to school.

Protected data now includes photographs, Social Security numbers, medical information, addresses, and other identifying information. Private investigators are granted access under certain circumstances, such as during a legal proceeding, to recover a debt, to uncover a fraud, or on behalf of the government or an insurer. Those who access such records are occasionally audited to ensure compliance.

The pickup truck, Foley found, was registered to someone named William Pras. Foley was an excellent researcher. Nearing retirement, she had a bachelor's degree in library and information science. She enjoyed crossword puzzles. Before becoming an investigator, she worked as a copy editor for custom-published magazines like those you find in your seat-back pocket on airplanes, and as an administrator at the New York Public Library.

Foley's skills were specific and useful. Although she had never tracked down a witness, never signed an affidavit, could not extract deleted photographs from a hard drive, could not speak Mandarin, and did not have a working knowledge of ballistics, she was obsessive and curious. She took to investigative work like a pilot or a surgeon, with a deep knowledge of her skill set and a methodical approach to crafting and adhering to checklists.

When searching for information, she drafted vast search strings with innumerable variations on the spellings of names and dates and then decanted her findings into clear pools of what intelligence officers call "actionable" facts. Foley was a pure practitioner, trained when information was found in books and written on typewriters or with Word Perfect.

While Gibbs wanted to pursue Spurn like a fugitive, Foley considered Spurn simply a puzzle to be solved. Younger investigators who know Gibbs were in awe of him and all of his spookiness, but they learned craft from Foley. She reminded me of Carol Loomis, whom *The New York Times* called "one of the country's most venerated financial journalists, with a collection of work that reads like a history of modern Wall Street." I worked with Loomis at *Fortune* in the late 1990s, early in my journalism career. She retired in 2014 after a sixty-year career, during which she was best known for befriending Warren Buffett in 1966 and writing one of the first articles about hedge funds.

One afternoon in 1999, as I was researching a story on political activism for *Fortune*, Loomis asked me about my article. We were in the magazine's often-deserted library, and when I confessed to having trouble finding a phone number for a government agency, she climbed atop a footstool and reached for the top shelf. Wiping away a layer of dust, she handed me a thick green tome with bright gold lettering.

The Green Book is the official directory of the City of New York. Published every year since 1918, it holds contact and background information on New York's city, state, and federal government, as well as county offices, the court system, and foreign consulates. Loomis didn't know it at the time, but that was just one class in an unofficial course she taught me in how to burrow for and locate elusive facts.

Spurn himself was not a defendant in 3-Draw's lawsuit, but Banning instructed us to treat Spurn as if he were indistinguishable from TopApp321. The opportunity to chase the assets of a company's owner was rare because of the way Top-App321 was formed—as a limited liability company, or LLC, an entity type that typically shields its shareholders from legal action. But the court ruled that Spurn's behavior had been so appalling (he showed contempt for the legal process by ignoring the judge's orders and stole money) that we could "pierce the corporate veil," which meant Spurn would be liable for Top-App321's debt.

Banning also instructed us to find assets in the names of Spurn's immediate family members. Foley ran dozens of data points through dozens of databases: Spurn's name, his wife's name, his addresses, his former employers, and companies he had formed, among many other identifiers. The databases included PACER (Public Access to Court Electronic Records), a valuable trove of official federal criminal, civil, bankruptcy, and appeals court records; Bloomberg Law; the Miami-Dade County Recorder's office; the Dallas Central Appraisal District; and the Delaware Division of Corporations. She looked for his golf scores, boats he owned, debts, web domains he registered.

Foley found articles of incorporation from companies Spurn formed, the vehicle identification numbers of his cars, tax filings from foundations where he was on the board, deeds from his real estate transactions, and complaints from lawsuits involving him.

She compiled an impressive dossier. Spurn had spent most of his career in Miami working for technology start-ups and investing in real estate, but for the past decade he had lived near Dallas. He owned the five-bedroom home in a new development that Gibbs was staking out. Spurn had a mortgage on the home, but otherwise no debts. He had a recreational hunting license issued by the Texas Parks and Wildlife Department. He was a registered Republican and had donated to the political campaigns of two state senate candidates. He played tennis on weekends at a private club and was on the board of a foundation that provided shelter to homeless teenagers. In short, so far Foley had found that Spurn lived an ordinary life and had few assets.

*A*s the jogger approached Gibbs, he wondered who she was and what she wanted. Maybe she was Spurn's wife. Maybe she was a neighbor who was aware of Spurn's frauds. Maybe she had already called the police. He steeled himself, but he wasn't worried. Gibbs spent his career deflecting probing questions from strangers.

"Hi," she said, bending down to pet Oscar.

"Hello," he said, bending down with her. They chatted about Oscar's name, breed, and age. They developed a quick rapport.

"You live around here?" he asked.

"Yes, not far."

"I'm trying to get to know the neighborhood. Curious about the real estate."

Suddenly the woman stood up and backed away. She was staring at his ankle: his pant leg was caught on top of a holster that held a pistol.

Every state regulates firearms differently, and some cities and counties have their own regulatory framework around guns. In New York, where Gibbs was from, anyone who carries a weapon must obtain a license and pass a background check. The license specifies the weapon's caliber, make, and model and the manufacturer's name and serial number, and it indicates if the holder is permitted to carry on her person or keep it in a particular place. In New York City, gun licenses are valid for three years. During recertification, the licensing body requires the license holder to submit her name, date of birth, gender, race, residential address, and Social Security number. She must also list any firearms she possesses and affirm she is not prohibited from possessing guns.

139 ■ THE MODERN DETECTIVE

In New York state, the name and address of gun license holders are public, but they can request this information be suppressed for a valid reason. For example, if the applicant's "life or safety" is endangered, if she is a "protected person" under an order of protection, if she is being threatened by a dangerous spouse, or if she is an active or retired police officer, peace officer, probation officer, parole officer, or corrections officer, according to the statute.

New York State Penal Law states that in New York City, to possess a gun, a license holder also needs a "special permit granting validity," which the New York police commissioner has wide discretion to issue. Gun owners don't need the special permit if they fall into a few categories: licensed gun owners who are passing through the city on a continuous trip with the gun in a locked container; licensed armored car security guards who are transporting money in an armored car during the course of employment; and retired police officers or federal law enforcement agents who already have a valid gun license.

As a retired FBI agent, Gibbs decided to play this one straight. He could not risk assuming a false identity to trick this jogger into turning over information, even useless information. He was not looking to entrap anyone. He was not seeking an indictment. He was simply looking for Jim Spurn's assets.

"Don't worry," he said, holding his hands up as if he were being threatened with a gun. "I'm a retired federal agent. If I don't wear it, I feel a kind of phantom limb syndrome. It's one of the perks, I guess."

She seemed relieved. She asked him about the job, which gave him the opportunity to talk about some exciting cases he had worked. Eventually, he steered the conversation back to her and real estate.

"How expensive is it out here?"

"Property values have soared."

Gibbs pointed at Spurn's house. "I'd buy that one," he said. "What would that run me?"

"A million?"

"Let's find out!" Gibbs started back toward Spurn's driveway.

"They're not home," she said.

"You know them?"

"Not really," she said. "They're not here very much. I think they spend a lot of time in New York City."

Other than Spurn's home, Foley had turned up few assets, but she was convinced he was hiding something. She turned her attention to Spurn's wife, Melinda, an attorney, who three years earlier retired as a county judge near Dallas.

One of the great lessons that Foley learned in her time as a PI is that juicy bits of information can be found in seemingly prosaic documents. For example, an affidavit to record a witness's memories of a burglary might include an elusive middle initial. A federal tax lien might show an unknown address. Federal Election Commission records might record where someone once worked. You are never sure where useful information might turn up.

Foley knew that judges must file financial disclosures that

contain the names of donors who contribute to their campaigns. She also knew that the Texas Ethics Commission kept personal financial statements, which require the disclosure of names, phone numbers, and addresses; the name of a spouse; all sources of income; mutual funds in which a judge, spouse, and dependent children invest; personal notes and lease agreements; interests in real property; and boards and executive positions, among other information.

These documents, which are designed to prevent conflicts between judges and litigants who appear in their courtrooms, are treasure troves of information for investigators whose clients are embroiled in commercial civil litigation.

The concept of judicial disclosure dates to the late 1960s, when Americans were mired in Vietnam, embroiled in the civil rights movement, and increasingly suspicious of authority, Charles Geyh, a law professor at Indiana University who specializes in judicial ethics, told me.

In 1969, Clement Haynsworth, a judge on the U.S. Court of Appeals for the Fourth Circuit, was nominated by President Nixon to the Supreme Court, but Haynsworth was rejected by the Senate after it was discovered he presided over cases in which he had a financial interest in some of the litigants that appeared before him. Although Haynsworth was not accused of any intentional corruption, his conflicts led to judicial codes of conflict and, ultimately, financial disclosure by judges at the state and federal levels.

Nearly two generations later, many judges bristle at the

burdensome requirements, they worry their privacy will be violated, and they fear aggression by disgruntled litigants, says Geyh.

But this disclosure, he adds, is consistent with the Fourteenth Amendment of the Constitution, which guarantees due process. Two hundred years into the American experiment, it occurred to active citizens, who long had access to varying degrees of disclosure from officials in other branches of government, that due process included the right to an impartial judge without financial conflicts. The Founding Fathers, it appears, did not appreciate the ability of Congress to regulate the judiciary.

"There may not be a constitutional right to disclosure," says Geyh, "but there is a constitutional dimension to ensure an impartial judge."

The disclosures are still lacking. According to an evaluation of state supreme court judicial reporting by the Center for Public Integrity in 2014, forty-two states and the District of Columbia received a failing grade, and no state earned an A or B, which meant most states had few processes in place for assessing judges' partiality.

Twelve states relied on "self-policing disciplinary bodies—made up of justices themselves—to enforce the courts' ethics rules." The center's investigation, which uncovered dozens of questionable gifts to judges, judges who authored opinions in favor of companies whose stock they owned, and other entanglements, did lead to some reforms, but nationwide there is still not full transparency.

This does not surprise Geyh. "Judges are naturally more gun-shy than legislators, who are more accustomed to scrutiny." In fact, until the first third of the twentieth century, judges' salaries were often determined by fines they assessed against litigants, so they were incentivized to punish people. Today, said Geyh, judicial financial disclosures have multiple purposes.

"The downside is they provide fodder for blackmailing judges. But they ensure litigants get a fair shake and that the public has confidence in its judges. And these disclosures have an ethical dimension in ensuring judges keep their noses clean and live up to the Platonic ideal of their profession."

Foley had other ideas for how to use the information. Two weeks into our investigation of Spurn, I asked her what she'd found.

"In Part 7B of Justice Melinda Spurn's Form PFS, called Interests in Business Entities, she listed a company called Prasit and disclosed she is a 50 percent owner."

"Is she an officer or director?"

"No."

"What does Prasit do?"

"It's not clear," she said, "but it may not matter." What it owns, she explained, is more interesting. She ran the name through the Texas Office of the Secretary of State SOSDirect database and found a company with this name that had a parent entity in Delaware. Both were formed in 2005. Nowhere in the records was a purpose of business stated, which is not uncommon, and Melinda's name was not on any of the documents. Next, Foley downloaded the articles of incorporation and all of the annual reports. And since 2011, the manager of Prasit has been William Pras.

Pras was the registrant of the pickup truck Gibbs saw in Spurn's driveway.

Foley dug into Prasit looking for news reports, Facebook posts, lawsuits, websites, and more and found Pras was sued in 2015 in Dallas County district court for breach of contract.

Court filings were not available online, so she dispatched someone, what investigators often call a "court runner," to manually obtain hard copies from the court archives.

Foley found that Pras was sued by a financial adviser seeking unpaid fees for setting up various investment vehicles for Prasit, Pras, and a handful of other companies and people, including Melinda Spurn. Pras and the other defendants, but not Melinda, countersued, alleging the investment adviser defrauded them.

"What do you make of that?" I asked Foley.

The claims in the lawsuit were complicated, she said, and much of it wasn't relevant for us, but there were clues.

According to the complaint, Pras and the other defendants simply failed to pay the adviser his fees, but Pras countered that the adviser was running a scheme in which he moved his clients' money among bank accounts in Panama without his clients' permission. The adviser then diverted a portion of those funds to accounts he controlled, which was a violation of his fiduciary duty and an obvious conflict.

In the filings was an affidavit by Pras's lawyer attesting to the general background of the case. The lawyer, Gene Gilles, who was based in New York, provided extraordinary details about the nature of their investments.

Gilles disclosed that the adviser formed a new entity called GATT1 LLC through which some of their investments—in golf courses, real estate, small cap equities—would flow.

Foley's excitement was palpable.

She told me that Gilles also mentioned, tangentially, that the adviser arranged for GATT1 to purchase fifteen thorough-bred horses that were used as collateral for a $2 million loan it sold to investors that promised guaranteed annual returns of 9 percent. Through this portfolio of equine assets, each investor owned a large fractional share of the horses.

Attached to the affidavit was a list of the investors. One of the names on the list was Melinda Spurn. Foley had stumbled on what looked like a trove of Melinda's money.

*I*n a 2014 essay on the mythology of Sherlock Holmes, the critic Laura Miller wrote in *Harper's Magazine*, "It is central to the pleasure of the Sherlock Holmes stories that they invite play, and that they were never meant to be taken seriously, either for the dastardly criminals whose designs provide the great detective with his cases or for the order restored to a snow-globe version of London once the cases are solved."

Real private detectives don't spend all their time catching dastardly criminals, and we certainly don't restore order to the universe.

And yet these are some of the characteristics that made

Holmes "arguably the most popular fictional character of the modern era," as Miller put it.

Since 2000, there have been a handful of real detectives whose own behavior has roused the interest of the media.

In 2008, Anthony Pellicano, whom the media enjoyed calling "the private eye to the stars," was convicted after two trials on dozens of charges, including racketeering, wire fraud, and wiretapping, after he bribed phone company employees to install software so he could eavesdrop on the calls of Hollywood celebrities and paid off law enforcement officers to obtain personal details about his targets, among other crimes. His clients included Michael Ovitz, the studio chief Brad Grey, the billionaire Kirk Kerkorian, and the comedian Chris Rock.

In a detail that could have been plucked from one of Pellicano's clients' scripts, the FBI began its investigation in 2002 after the *Los Angeles Times* reporter Anita Busch found a dead fish with a rose in its mouth on the hood of her car along with a sign that read "Stop." Busch later testified she blamed Pellicano and Ovitz for the threat after she co-wrote articles about Ovitz's financial troubles and a possible sale of his talent agency. Pellicano was released from prison in San Pedro, California, in March 2019.

In 2005, Patricia Dunn, the chairwoman of Hewlett-Packard, instructed her company's lawyers at Wilson Sonsini to hire an investigative firm called Security Outsourcing Solutions to help figure out if any HP board members had released private information to a news reporter without authorization. (Outsourcing Solutions subcontracted some of its work to another PI firm called Action Research Group. This was not the first time Hewlett-Packard had hired Security Outsourcing Solutions.) Earlier that year, executives at HP were alarmed by details about a company retreat that had shown up in news reports that could only have come from its directors, who leaked information to the media.

According to government records, Action Research Group targeted HP board members, employees, others "affiliated with HP," HP employees' family members, reporters for *CNET, The Wall Street Journal, The New York Times,* and *BusinessWeek*, and family members of the reporters to try to learn where the leak came from.

The investigators dug up names, addresses, phone numbers, dates of birth, Social Security numbers, mobile billing records—mostly private and confidential information. Among the tactics the investigators used was to assume the identities of their targets to obtain these records. The investigators also used the subjects' Social Security numbers without their knowledge to get their phone records.

When the tactics of the HP investigators became public, outrage ensued, in the form of media coverage, as did criminal and regulatory investigations, including one by Congress. In 2007, a federal bill was signed into law making it illegal for any private individual to use false pretenses to buy, sell, or obtain another individual's phone records without permission, a tactic known as pretexting. (Many private investigators still pretext. In fact, we are sometimes hired to catch other such nefarious PIs.)

The only person who served prison time for this security

breach was Bryan Wagner, a relatively junior investigator who swindled phone company representatives into giving him records of HP directors, journalists, and their families. He pleaded guilty in 2007 to "aggravated identity theft" and spent three months in prison. Two of his colleagues received probation.

Charges against the HP executives were dropped or were thrown out by a judge. (In Dunn's case, the charges were dropped in part because she was in the late stages of breast cancer; she died in 2011.)

*W*ith the tip from the jogger that Spurn spent time in New York City and the discovery of Pras's truck in Spurn's driveway, Foley looked for links between Spurn and Pras and for clues that Pras had a New York footprint. She found one: a two-bedroom apartment on the Upper East Side that was owned by Pras's wife.

The following week, Gibbs and his team set up on surveillance outside Pras's apartment building in a rental car. (While they waited, they listened to Howard Stern.) Their hope was to catch Pras with Spurn. Were the two meeting with business partners? Renting office space? Buying real estate?

During their second day of surveillance, Gibbs followed Pras to a midtown hotel where Gibbs knew someone on the hotel's security team.

Because of the pension policies in city police departments, many police officers retire young—some in their early forties, if they join the force a few years after high school—and have second careers in related fields: as armed guards, as private investigators, as corporate security officers, or doing executive protection. Some eventually rotate back into government service to help politicians manage law enforcement agencies or run security detail.

For example, William Bratton began his career as a Boston police officer in 1970, became chief of the New York City Transit Police in 1990, was appointed commissioner of the New York Police Department in 1994, joined Kroll as a private consultant in the early years of the twenty-first century, moved to Los Angeles to become chief of police in 2002, and considered joining the British government as an adviser. In more recent years, he formed Bratton Technologies Inc. and the Bratton Group, joined Altegrity Risk International as chairman, worked as an analyst for NBC News, and returned to New York City again in 2014 as police commissioner. In 2016, he left the NYPD and joined Teneo, a company that advises corporate executives. He is also the vice-chairman of the Homeland Security Advisory Council.

This is law enforcement's revolving door, and it is one reason investigative firms staffed with private eyes from different fields are effective: they have deep networks.

My background is in journalism, so I know reporters and editors around the world, including those who left the writing life to join crisis management and communications firms or became publicists or government policy advisers. Foley's friends were in academia. As a result of his contacts, it took Gibbs one phone call to get Spurn's hotel room number. (Of course, that information should not have been given out, but it can never hurt to ask.)

The next day, Gibbs and his team were in the hotel itself setting up in the room directly across the hall from Spurn. When they checked in, their only luggage was two Pelican cases filled with listening devices and video cameras. They slipped a small camera under their door so they could watch Spurn come and go. Gibbs posted four other investigators, two to a car, down the block from the hotel.

The next morning, Spurn left his room and headed out to catch a cab. Gibbs's men saw him leave the hotel, and when he approached the curb for the doorman to hail him a taxi, they started their engines and communicated using two-way radios so as not to leave a texting or email trail. Gibbs's team followed Spurn's taxi to Café Centro, a sprawling restaurant just north of Grand Central Terminal. Gibbs followed Spurn into the restaurant while his team parked the cars in nearby lots.

On any given weekday in midtown Manhattan, there are, I would estimate, dozens of surveillance teams shadowing people. Gibbs had followed targets to this very restaurant, and he knew how much surveillance takes place around Grand Central because of the concentration of restaurants and hotel bars near the train station that cater to people having affairs. It's a convenient neighborhood for such behavior, and the hordes of commuters and tourists make it easy to blend in. Gibbs has followed men on three separate cases into the same Irish pub on Forty-third Street that he calls a favorite for philanderers.

Spurn was no cheater. He was a thief who'd absconded with $9 million belonging to 3-Draw. Gibbs tipped the maître d' at Café Centro and asked for a booth near Spurn, who was seated with two other men. When two of Gibbs's team members arrived, the three of them pretended to be friends reuniting after many years. They hugged and took photographs, including fake selfies that not so surreptitiously captured the Spurn table. Then they settled into a meal near Spurn and eavesdropped.

Spurn's conversation was banal and provided no new intelligence, but before the checks were paid, Gibbs emailed the photographs to Banning, who forwarded them to his client, who recognized one of Spurn's guests as one of the original investors in TopApp321 who had met with the 3-Draw team on a few occasions to discuss app development. Spurn, it seemed, was planning something new with at least one of his old cronies.

After tailing Spurn back to his hotel, Gibbs left his surveillance team on Spurn and returned to the office to meet with Foley.

A proper asset search involves not only finding assets but valuing them accurately. For instance, a penthouse in Palm Beach that was bought with a down payment of $200,000 and a loan of $1.8 million from Citibank is worth only $200,000 to the owner, because the rest of the money is Citibank's. I once tracked a wine collector who drove a fleet of yellow Ferraris, only to discover they were leased. I know a private eye who found a room full of Mondrians that turned out to be forgeries.

At the office, Foley called a few horse breeders and others who understood the industry and learned that while horses are risky investments, they can hold value. If the horses are fast, if they race, and if they race and win, they make money. Races are assigned different grades. Stakes races are the most profitable and prestigious. And there are a few markets outside the United States, like Dubai and Australia, where owning thoroughbreds can be profitable as well. Also, a winning horse may have a breeding career: there is money in stallion sperm. And horses have resale value. A stallion becomes a sire, while a dam, a female, gives birth to foals, which may grow up to become racehorses.

Profits from races, breeding, or sales would be distributed to the investors—like Melinda Spurn.

Foley, Gibbs, and I met the next day to finish our conversation. Foley explained that in the last paragraph of Gilles's affidavit, he says the adviser had been traveling to Mexico for treatment for an unspecified illness and failed to alert his investors to changes in his investment strategy and decisions.

Gibbs understood what she was getting at. "Spurn and his wife owned shares in thoroughbreds worth millions." He thought for a second. Then his eyes widened.

Foley finished the thought for him: "And they don't even know it." This was of no relevance to the investigation (money is money is money), but it was an oddity of the case that we'd found assets of which their owners were not aware.

Gibbs's and Foley's complementary investigative skills were both necessary in helping Banning to recover his client's money, which he did when he seized accounts in Spurn's name (we played no role in that part of the legal proceedings). As a researcher, Foley is a consummate rule follower. Gibbs, however, who carries a weapon even as a private citizen, is often nostalgic for his time in law enforcement, when he was granted permission to break rules the rest of us must follow.

Much is made of the dark side of private intelligence gathering. In his 2010 book, ominously titled *Broker, Trader, Lawyer, Spy: The Secret World of Corporate Espionage*, the journalist Eamon Javers takes a dark view of my industry by cherry-picking dastardly acts by international men of mystery.

He recounts how a British company called Diligence, founded by a retired MI5 officer, was caught running a covert operation to infiltrate the offices of the accounting firm KPMG, on behalf of a Washington lobbying firm that was retained by a Russian oligarch.

Javers also reports how an anonymous financier—who was enthralled with an Israeli model named Bar Refaeli who was then dating Leonardo DiCaprio—hired a "spy firm" to conduct surveillance on DiCaprio in South Africa while he filmed *Blood Diamond* in 2006.

The "ultimate prize" would have been a photograph of DiCaprio "in the arms of anybody other than Bar Refaeli, which could be leaked to the media or mailed to Bar. Presumably, if she was confronted with evidence of DiCaprio's treachery, Bar would be more easily lured into the waiting arms of the financier," writes Javers, who concedes the story may not be true.

These are thrilling, if rare, tales. In fact, investigators in the

United States are usually employed by lawyers, including the world's savviest and most ethical. When we work with lawyers who are in the middle of a legal dispute or who are working in anticipation of litigation (or to aid them in communicating with clients), our work product and our communications with the lawyers are privileged and confidential and can't be shared in open court or anywhere else without the permission of the client and the lawyer. Private eyes, who become agents of the lawyers, must respect the rules of the legal community. In these cases, we essentially step into the shoes of the lawyers who engage us.

Private detectives are also often hired to collect facts designed to be made public and to be used in criminal and civil legal proceedings—in other words, to root out wrongdoing that, arguably, benefits society.

We help nonprofits uncover expense account fraud. We help legislative committees investigate government corruption. We help bankruptcy trustees recover missing money.

In 2009, Kroll worked with the City of New York to run an undercover investigation into the illegal sales of firearms at gun shows around the country, because it was tracing the flow of firearms into New York. According to a Kroll report, which was made public, the company uncovered damning information that the public needs to know: "Seventy four percent of those approached by Kroll completed sales to people who appeared to be criminals or straw purchasers." Information gathered during the stings was used by the then New York City mayor, Michael Bloomberg, to help shape public policy in an attempt to get illegal guns off the city's streets.

In 2013 and 2014, the Mintz Group (where I worked for six years) spent months collecting documents and interviewing dozens of witnesses on behalf of a special investigative committee of the Utah House of Representatives to probe allegations that the then Utah attorney general, John Swallow, "compromised the principles and integrity of the Office to benefit himself and his political supporters," according to a report on the committee's findings. The report concluded that Swallow "hung a veritable 'for sale' sign on the Office door that invited moneyed interests to seek special treatment and favors." Swallow resigned in November 2013, four months before the report was made public.

Gibbs was discreet, but he did not break any rules in chasing down Spurn's assets, even though, I suspect, he wished he could have.

Another former government employee, John Rowland, who was the governor of Connecticut, was not as restrained as Gibbs, and his refusal to behave ethically had devastating consequences.

8.

A CIGAR, A COOKIE,
AND A CANOE

Not far from the State Capitol in Hartford, Connecticut, there is a dark bar on a country road where locals drink hard liquor. In early 2004, a private eye named Jim Mintz met a state political operative at this roadhouse. Over tumblers of bourbon, they discussed John G. Rowland, who was then the governor of Connecticut.

Little more than a year earlier, when Rowland had ended his term as chairman of the Republican Governors Association, his colleagues and allies rewarded him with "nearly universal acclamation and admiration," according to the editorial board of the *Hartford Courant*.

The paper predicted Rowland could "climb higher up the party ladder, perhaps as a Cabinet member, U.S. senator, party chairman, or beyond." Rowland "would seem to have all the requirements for advancement": youth (he was in his mid-forties), a string of campaign successes (he had recently won his third gubernatorial term), connections (he was "a pal of the president," George W. Bush), and charisma ("he schmoozes effortlessly with the national media"). Rowland "will be whispered as a possible vice presidential candidate . . . and maybe even for the top."

Mintz's source, however, was less optimistic about Rowland's future. Local media had recently reported on the governor's inappropriate relationships with building contractors, and Mintz's source was concerned for the GOP's leadership, because the source was himself a Republican. The source had agreed to meet with Mintz, who was investigating corruption in the executive branch of the Constitution State. They met every couple of weeks, and Mintz would bounce ideas off him about witnesses he might contact or theories about the governor's conduct.

Mintz worked hard to prove to his source that he was not looking to hang anybody. He just wanted the facts about Rowland's conduct.

*G*athering facts is what Mintz has spent his professional life doing and what he did for four frantic months in the spring of 2004 when his corporate investigations firm, the Mintz Group, was retained by the bipartisan Select Committee of Inquiry of the Connecticut House of Representatives through its special counsel, Steven F. Reich, then a partner at the national law firm Manatt, Phelps & Phillips.

Reich had experience handling government investigations. He advised U.S. House Judiciary Committee Democrats during the impeachment proceedings of President Bill Clinton in 1998, and as senior associate White House counsel in the 1990s he defended senior officials from congressional and Justice Department investigations.

In late 2003, the *Hartford Courant* published credible articles about how Rowland accepted money and favors, possibly in return for state contracts or other benefits. Influence peddling in Connecticut was front-page news across the state. In November 2003, one report questioned the source of payments for renovations to a vacation cottage Rowland owned on Bantam Lake in Litchfield, Connecticut.

At first, Rowland responded to the media reports by saying he and his wife, Patricia, paid for the repairs. However, in the weeks that followed, it became clear that some contractors were not paid until years after their work was completed. When faced with this proof, Rowland, according to the media, admitted to lying about the source of the payments, apologized, and disclosed that friends, members of his staff, state employees, and state contractors paid for the work or provided it for free. But he refuted allegations that anyone received a state benefit in return.

His honesty play failed. Calls for a further accounting rose to a fever pitch, especially following similar reports questioning the propriety of Rowland's lease and sale of a condominium in Washington, D.C.; other gifts he received, including a hot tub; and vacations he took.

The Connecticut House of Representatives formed the committee to investigate and recommend "whether sufficient grounds existed to impeach" the governor, according to a report it later issued. It was not a prosecutorial body; it had the authority to collect and present evidence but not to indict, try, or convict.

Mintz and Reich had never met, but Reich knew something about private investigations (he had recently married a private eye unconnected to Mintz's firm). He cold-called Mintz and told him he needed a seasoned investigator in Connecticut.

The Mintz Group is based in New York and, at the time, had offices in Washington, D.C., Miami, Chicago, San Francisco, Hong Kong, and London; it has since expanded to India and Canada, among other places. Mintz had few contacts in Connecticut, but he landed the case anyway. It was, perhaps, his lack of any local conflicts that made him a shrewd choice, because he could approach the job with an unbiased eye.

Mintz and Reich knew they were wading into well-trod territory: there was a concurrent federal criminal investigation of the governor and a number of his associates.

The next day, Reich and Mintz traveled to Hartford to meet the legislative committee, which agreed to hire the Mintz Group to work under Reich's direction. Mintz started working quickly, and he was soon drinking bourbon in the roadhouse near Hartford with a secret source.

I used to work for Mintz, whose voice booms with self-assurance. He is a preacher as much as an investigator. (He teaches investigative techniques at Columbia Journalism School.) When you are interviewing someone, he would remind us, try not to ask questions. Instead, form relationships, gain sympathies, ask for help. Using that strategy, Mintz won over his source.

One evening while meeting with Mintz, the source mentioned that the governor sometimes flew on an air charter service called Key Air. The source said Rowland and Key Air appeared to have a close relationship.

Mintz did not give much credence to this lead. He'd never heard of Key Air, the media had not covered it, and his team had not considered this angle. However, out of curiosity, Mintz sent one of his investigators, Suzanne Clarke, a former journalist, to the Key Air office at Waterbury-Oxford Airport, which is about forty miles southwest of Hartford. Mintz's assignment for her was simple: ask them if they would talk about their work for the governor.

Clarke remembers the day well. "As I'm driving, I'm getting nervous," she told me. "It's the sort of nervousness that gives me adrenaline. I was processing what I was going to say primarily to make sure that I'm not being deceptive. A lot of people have this idea that investigators go into those types of situations with a scheme, a disguise, and they coax the information out of somebody by misleading them. That's exactly the opposite of what happens or should happen."

Not long after she arrived at Key Air's office at the airport, Clarke called Mintz.

"They slammed the door in my face," she told him.

Mintz was shocked. He asked her to tell him exactly what happened.

"I knocked on the door and said, 'Hi, I'm Suzanne Clarke. I am an investigator with the Connecticut Select Committee of Inquiry,' and I explained why I was there. But they wouldn't answer any questions. They shut me down."

Mintz called Reich, told him Clarke's story, and said he wanted to play a hunch. Mintz wondered why Key Air would be so evasive. Perhaps his roadhouse source was onto something. Mintz wanted to subpoena Key Air for documents related to Rowland and his office. He knew Key Air would have to respond to the subpoena within twenty days.

Reich was intrigued, but he was hardly excited by the lead. Dates for a hearing had been set. The investigation felt like a race against the clock. After some pleading on Mintz's part, Reich granted his request for a Key Air subpoena.

In 2004, the Mintz Group was a decade old and employed about seventy-five investigators, including Andrew Melnick, a former prosecutor from the District Attorney's Office in Manhattan and general counsel of the New York City Department of Investigation, who played a significant role in the Rowland investigation.

But for all Mintz's manpower and expertise, Mintz's team made little progress during the first month of work and developed few leads. A few witnesses had spoken off the record, but most of the people he interviewed clammed up. Reich feared the little evidence they had was useless because it was hearsay, and therefore not admissible in court.

The pressure was on Mintz. Many of the witnesses they approached retained lawyers and declined to speak. Some had been interviewed by the FBI and were loath to answer the same questions again. Some sources, such as members of Rowland's security team, who were law enforcement officers, were conspicuously unavailable.

"Those troopers were in the wind," Mintz later told the *Hartford Courant.* "Those guys know how to duck an investigator." (He was referring to troopers who were asked to speak voluntarily, not witnesses who were compelled to speak to the committee.)

Other witnesses invoked their constitutional right against self-incrimination or failed to respond to the committee's demands to turn over documents. At one point early in their investigation, Mintz and Reich met with Nora Dannehy and Eric Glover, prosecutors from the U.S. Attorney's Office in Connecticut, which was also investigating Rowland, hoping to develop a collaborative relationship. They were rebuffed.

Then Mintz had an idea. He had recently interviewed Richard Mulready, a former member of the board of the Connecticut Development Authority, or CDA, a now-defunct agency that financed struggling companies in the state. Mintz knew that in December 1996, Fabricated Metal Products, which was owned by Robert Matthews, a government contractor and a friend of Rowland's, applied to the CDA for a $6.8 million loan guarantee.

Earlier that month, the CDA delayed reviewing FMP's application amid "numerous negative comments" from board members, according to the committee's report; Mulready, for example, was planning to vote against the proposal because of concerns he had about Matthews's financial condition. The board knew Matthews had a "history of taking cash out of his companies," but it nevertheless considered a revised proposal in January 1997.

Among the nine board members at the January meeting were the CDA chairman, Arthur Diedrick, Paul Silvester, and Mulready. Mintz knew the board was on the verge of voting against the proposal when a recess was called. After the recess, the proposal passed by a single vote. But it wasn't clear why it passed at all.

During his interview with Mulready, Mintz asked whether Rowland had influenced the vote. Mulready said he had not.

Somebody, Mintz realized, wasn't telling the truth.

*M*intz formed his firm in 1994 after parting with Terry Lenzner, with whom he ran IGI, a detective firm in Washington, D.C., where Mintz grew up. (Lenzner had been assistant chief counsel to the Senate Watergate Committee in the 1970s, and his bio on the IGI website states he "authored and personally served the first Congressional subpoena of a sitting U.S. President.")

Mintz has roots in Washington. His father, Seymour Mintz, was a lauded tax lawyer who had been an intelligence officer in the navy, worked in the IRS's legal department, counted Howard Hughes among his clients, and helped run the law firm that is now called Hogan Lovells. Mintz's brother John Mintz, who works at the Mintz Group, spent more than two decades as an investigative reporter for *The Washington Post* covering terrorism, security, campaign finance, and defense.

Mintz has a corner office at his company's headquarters in New York on Fifth Avenue near Union Square. On his office wall hang framed historical documents that symbolize his insistence on obtaining primary sources to solve cases. He has a framed quotation from a client that reads "This is a great memo, but it's too factually based."

Mintz often talks about the process of information and intelligence gathering. I have heard him say that every complex investigation involves digging into a series of events that happened in the past: what happened over there, in that place, at that time. Mintz often approaches cases by considering who was in the inner circle, finding those witnesses, and getting them to talk to him. He took that approach in the Rowland case.

While investigating the CDA vote on FMP's loan application, Mintz found one name in the record intriguing: Paul Silvester, a former investment banker who in 1997 was appointed Connecticut state treasurer by Rowland. At the time, Silvester was the youngest person in the country to have that job.

But soon his career careened. In November 2003, he was sentenced to nearly five years in prison for taking bribes in return for placing more than $500 million from the state pension fund in private investment funds.

The media reported ties between Rowland and Silvester and suggested Rowland was aware of, and even abetted,

Silvester's crimes: Silvester helped raise campaign funds for Rowland, and the two appeared at fund-raisers arranged by Silvester in Hartford and New York City that were attended by money managers who received, or would soon receive, money from the state pension funds to invest.

Rowland denied knowing about Silvester's scheme, but Mintz felt Silvester might talk about Rowland. So he asked for official permission to approach him in prison. Reich did not object to Mintz's proposal, but the committee refused.

For several weeks, Mintz made only incremental progress on the case, though he corroborated some of what the media reported, including that the governor had helped steer state contracts to people and companies that made renovations, or paid for renovations, to his cottage in Litchfield, which, the committee eventually found, included building a patio and installing a furnace. But that kind of evidence was not impeachment material. Mintz worried he would be called to testify and have nothing substantive to say.

Mintz again asked the committee for permission to interview Silvester, and eventually it agreed. It felt like a desperation play.

Mintz arranged the interview through intermediaries, including Silvester's lawyer, and although he hoped to meet with Silvester in person at Eglin Air Force Base near Pensacola, Florida, where he was imprisoned, after a few weeks of wrangling, it was decided that the conversation would take place over the telephone.

The atmospherics were terrible. The committee was thirty

days away from the hearing, and everybody was skeptical that Silvester would have anything valuable to say. Then word came that the Eglin warden had offered the phone in his office to Silvester. Mintz and Melnick dialed the number.

After brief introductions, Silvester interrupted the conversation.

"Do you have a pen handy?" he asked. "Would you mind if I just talked?"

An hour later, the warden ended the call.

Silvester knew what Mintz wanted, and without being asked, he gave it to him. (Perhaps Silvester was feeling penitent.) According to the committee's report, which Mintz helped draft and which drew, in part, on his interview with Silvester, "Silvester confirmed that there had been a break in the [CDA] meeting, but said that, rather than receiving a telephone call from the Governor or someone in the Governor's Office, he and Mr. Mulready stepped aside for a private conversation with Mr. Diedrick, who conveyed to them both that the Governor desired to see this proposal pass."

Silvester said he asked for the aside because, according to the report, "during the Board discussion, Mr. Diedrick had commented that Mr. Matthews was a friend of the Governor's and Mr. Silvester thought the comment was inappropriate."

As a result of what he'd learned in his phone call with Silvester, Mintz reinterviewed Mulready and asked him to respond to Silvester's version of events.

This time, Mulready changed his story. He recalled that "the three men discussed the fact that the Governor wanted this deal done," the committee's report states. "Messrs. Diedrick and Silvester then each indicated that they would abstain and Mr. Mulready indicated that he would vote 'yes' to accommodate the Governor."

According to the committee's final report, Mulready "substantially corroborated Mr. Silvester's story and agreed to submit to the Committee an affidavit describing his recollection."

Diedrick testified that he did not recall whether a recess was taken or whether he had a private conversation with Silvester and Mulready. He said that he "did not recall whether anyone mentioned that Robert Matthews was a friend of the Governor or that the Governor had an interest in FMP's application."

Diedrick also testified that "he did not recall abstaining from the vote and, thus, could not explain why he abstained," and although he recalled speaking with David O'Leary, then the governor's chief of staff, about FMP, he could "not recall when that conversation took place or what was said."

The other six board members told Mintz's team "that they did not recall anyone mentioning that Mr. Matthews was a friend of Governor Rowland or that the FMP proposal was important to the Governor. None of them recalled a recess being taken, some noting that a recess would have been highly unusual, and some noting that 'breaks' were not usual."

The CDA board approved the FMP proposal. Mulready's "yes" vote proved decisive. On March 27, 1997, FMP obtained $16 million in financing from Norwest Business Credit Inc., conditioned on the CDA guarantee, which by then had been reduced to $3.6 million.

It is not uncommon for witnesses like Silvester, who have damaged reputations, to be truthful after the fact and for their testimony to be valuable even if their credibility is not. When bad consequences, or blame, rain down on an organization and reputations and lives are ruined, some people feel that blame is not distributed fairly. Good people get hurt; bad people get away with things. These circumstances loosen tongues.

In early 2018, my firm worked tirelessly for the rapper Meek Mill, who had been imprisoned in Pennsylvania for minor probation violations. Among the most valuable evidence we obtained—which inspired, in July 2019, a state appeals court to overturn Meek's underlying conviction—was the testimony, in the form of two sworn affidavits, by disgraced former police officers who provided credible evidence that Meek's arresting officer was corrupt.

In the Rowland case, Mintz had made other discoveries. Among them was that Rowland failed to prevent William Tomasso, who ran a construction and property management business, from gaining access to nonpublic government information in advance of the issuance of an RFP, or request for proposal, to build the Connecticut Juvenile Training School, or CJTS, in Middletown.

In addition to Tomasso's learning about the state's plans before any of his competitors, the media later reported that Tomasso won the award through a "fast-track" process by bribing Rowland's co-chief of staff, Peter N. Ellef, with cash and gold coins. (Both Tomasso and Ellef were later jailed for bribery and tax fraud.)

While digging into Rowland's links to CJTS, Mintz gained the cooperation of the Connecticut Attorney General's Office. Although the AG's Office was law enforcement and Mintz was not, both were part of the state structure, Mintz temporarily during his work for the committee. They shared an investigative ethos.

During the case, Mintz and his team also interviewed two of Rowland's aides. Christine Corey, Rowland's executive assistant, told committee investigators that when Rowland learned that Tomasso and Ellef were in Ohio together looking at a model facility in advance of the CJTS contract award, "the governor was very upset." The Ohio trip gave Tomasso a competitive advantage. The committee "found no evidence that Rowland confronted Ellef or Tomasso about the trip . . . or that he took any other action to remedy the situation."

In speaking to witnesses, investigators sometimes think they're looking for whistle-blowers. In the Rowland case, Mintz cast a wider net. Indeed, he didn't care whether a witness was aggrieved by Governor Rowland or was an admirer. Mintz wasn't looking for quotes; he was looking for names, dates, and amounts.

The stories that Rowland's assistants told Mintz's investigators—including that Rowland was unaware of Tomasso's trip to Ohio—were designed to be neither harmful nor beneficial to the governor; they were simply truthful, according to Mintz. It is possible they believed what they said would exonerate the governor. In fact, their statements were not exculpatory.

In August 2005, Jodi Rell, who would succeed Rowland as governor, closed the CJTS, which was designed to hold 240 boys convicted of nonviolent crimes. She cited mounting costs, harsh treatment of inmates, and outmoded approaches to criminal justice reform as reasons. *The New York Times* reported that the CJTS became a "symbol" of Rowland's "folly and corruption."

Today, when Mintz hosts guests in his New York office, he often points to a framed article from the *Hartford Courant* about Rell's decision to shutter CJTS. "This happened as a result of some investigative work I did," he tells guests.

Some of that investigative work was done by Suzanne Clarke, the Mintz Group employee who was rebuffed by Key Air at Waterbury-Oxford Airport. Clarke was hired by Mintz because she excelled at persuading people to talk. After a career in journalism—at *20/20*, *PBS NewsHour*, and a nonprofit investigative news organization called the Public Education Center—she joined Mintz in the late 1990s.

She was skeptical of the industry. "I had this perception that private eyes were stealing documents out of trash cans doing icky political stuff," she said. A headhunter courted her and persuaded her that the Mintz Group was not that kind of firm.

She quickly proved adept. Clarke loved to skateboard, and she skated around Washington, D.C., where she grew up. One day, she fell off her board and broke her arm. While recovering, she worked on behalf of plaintiffs in a class action lawsuit that required her to interview witnesses. Wearing a cast, Clarke showed up at a witness's home. The witness "took pity on me because I had a cast on all the way up to my shoulder," said Clarke. The witness even took notes for her, because Clarke could not write, and she drew a picture of an office of her former employer that showed the location of a ledger hidden in a filing cabinet.

During the Rowland investigation, Clarke's job was to show up unannounced at witnesses' homes or places of work and get them to talk. One focus of the probe was real estate transactions involving the governor, including a 465-square-foot condominium in the Capitol Hill neighborhood of Washington, D.C., that he bought in June 1989 for $57,500. According to Mintz, the media had done excellent reporting to question the source and amount of rental income the governor received from the condominium, its eventual sale, and his ties to Robert Matthews.

Rowland might have tried to rent the apartment in 1995: a draft of a newspaper advertisement for the property from Rowland's personal files had black marks at the bottom of the page that "obliterated some writing underneath," according to a forensic document examiner retained by the committee who determined the writing was "$600/mo" and "$50,000." (The committee found no evidence that Rowland rented or sold the apartment in 1995.)

About six months later, in March 1996, Rowland met with Matthews and Matthews's attorney in the governor's office, and later that day Matthews sent a letter to his niece, Kelly

Matthews, "indicating his intention to provide her with a monthly gift of $1,750 to cover her rent in Washington, D.C." Every month for the next two years, Matthews wired $1,750 into Kelly's bank account, and Kelly paid Rowland that amount.

The committee's report was careful not to draw explicit links. Instead it stated, "On May 15, 1996, Governor Rowland appointed Gerry Matthews, Robert Matthews's brother and a licensed real estate broker, to the Connecticut Real Estate Commission. The position is unpaid."

A year later, Matthews asked Wayne Pratt, an antiques dealer who also appraised items on PBS's *Antiques Roadshow*, to buy Rowland's apartment. Matthews told Pratt he could not buy it, because he was "too personally involved," according to the committee report. Matthews fronted the money and said he would reimburse Pratt for any losses. Matthews also asked Pratt "to disguise the nature and use of the funds" within Pratt's business records.

In May 1997, Matthews gave Pratt two checks made out to Wayne Pratt Inc., and in June 1997, using that money, Pratt bought the condo from Rowland for $68,500. (Pratt never visited the condo, according to committee records.) He paid another $5,000 for furniture. After paying closing costs and the mortgage balance, Rowland made more than $20,000 on the sale.

In December 1999, Pratt sold the property for $37,500; Pratt's office manager calculated that Matthews owed Pratt about $22,000 "for losses incurred in connection with the condominium." Matthews requested that a "reconciliation payment" flow through the records of Pratt's antiques business to "disguise the true purpose of the payment." Matthews also asked Pratt and Pratt's office manager to destroy any paperwork connected to the reconciliation. No records were destroyed. (Pratt and his office manager provided affidavits to the committee.)

Reich's investigative team had high burdens of proof. They sought to bring evidence to the Connecticut House committee showing not just possible corruption (that Rowland received something from Matthews, that Matthews received something from Rowland, that the deal was funneled through a front man) but that any transactions were unusual and occurred outside any competitive bidding processes.

Mintz asked Clarke to find an expert to testify about condominium sales in the Capitol Hill neighborhood of Washington in 1997. Although some members of Congress lived in the area because of its proximity to their offices, it did not at the time generally attract wealthy residents, particularly governors from faraway states. Clarke asked colleagues and friends for recommendations, and the responses were consistent. "You're not going to get a D.C. realtor to spill it on a public official," realtors told Clarke. "That's bad business."

After weeks of rejections, Clarke finally tracked down a promising lead: Pamela Kristof, an agent with more than twenty years of experience in Capitol Hill.

When Clarke asked Kristof if she would testify, she showed tepid interest. But Clarke remained optimistic. "When I told her the condo was sold in 1997 for about $68,000, she laughed out loud"—because the price was absurdly high.

With the date of the committee's hearing approaching, Reich's team was desperate to lock down an expert, so Clarke lobbied Kristof hard. Eventually, Kristof agreed and she and Clarke flew to Hartford. Meanwhile, other Mintz Group investigators researched sales of comparable apartments in the neighborhood. They hoped the data would support the testimony. Kristof's role, explained Clarke, was to bring color to the ridiculousness of the sale price to prove it wasn't a fluke— that it was, in fact, manipulated.

In her testimony, Kristof agreed that Rowland's price tag of $68,500 "did not reflect the fair market value of his condominium as of the date of the sale," according to the committee report. Kristof testified that the fair market value was "$25,000, maybe, maybe on a good day, $30,000." She also provided sales figures supporting her claims. For example, a unit identical to Rowland's sold on June 9, 1999, for $40,000 after having been on the market for 695 days, with an asking price of $42,900.

*T*he committee's hearings, which began on June 8, 2004, were held in the Old Judiciary Room of the State Capitol, a building that had been designed by a cathedral architect in the High Victorian Gothic style and erected in the late nineteenth century of marble and granite with an imposing gold-leaf dome. Among those who testified were Andrew Melnick of the Mintz Group; Martin Lippe, a forensic accountant; and

Rachel Rubin, the former managing director of the State Ethics Commission.

On June 15, Mintz took the stand for the second day. At the time, *The New York Times* had been reporting that, presumably because of the mounting pressure on him due to his public testimony, rumors were swirling that Rowland's legal team was seeking some kind of an agreement with the committee that would include the governor's resignation. Ross H. Garber, one of Rowland's lawyers, refuted the rumors. Rowland, he said, "is going through this process to the end, where the committee will have to decide if there is clear and convincing evidence of impeachable conduct."

Mintz was eating lunch in his temporary office upstairs in the capitol when he was handed a package containing a few dozen documents with a cover letter from Key Air's lawyer, explaining that the records were provided in response to a subpoena.

Mintz spread the documents on the table and arranged them chronologically. They showed Rowland took five trips on Key Air during the past few years. Each trip was documented with a flight manifest, a log showing departure and arrival dates, times, and airports, as well as an invoice and a check from the State of Connecticut or a Republican organization that corresponded to each invoice. Four of the five trips looked innocuous. But for one trip, to Las Vegas, there was no invoice and no check.

Meanwhile, Mintz's team scrambled to corroborate the Key Air discovery. They found photographs of the plane on Key Air's website. They checked Rowland's American Express card charges and his public and private calendars. His public calendar for November 15 and 16 showed no mention of Las Vegas and a notation "Keep day open per the governor."

Mintz called the Key Air lawyer who had sent the documents.

"I want to tell you what I'm seeing, and you correct me if I'm wrong," he began. "I see five trips taken but only four trips invoiced. The uninvoiced trip was to Las Vegas—never invoiced, no check."

"Where in your subpoena does it say I have to talk to you?" the lawyer asked.

"That's your position?" Mintz said.

"That's my position."

"By the way, I am testifying right now on live television."

"Yes, we're aware of that. We're watching you."

"Well," said Mintz, "I'm telling you that it looks like Key Air gave the governor a free trip to Las Vegas with his buddies."

"I'm going to repeat, Mr. Mintz: Where in your subpoena does it require me to talk to you?"

"Okay, I get it now. You've said it twice. Thanks very much."

Mintz returned to the witness table, and he and Reich wrapped up a line of questioning about vacations Rowland took that were paid for personally by Peter Ellef, his former co-chief of staff. Reich then turned to Key Air.

"Mr. Mintz, was part of the investigation that you conducted relating to the 'gifts issue' centered around a flight company called Key Air Inc.?"

"Yes."

"If you could lean into the mic, please, and just very briefly give us an overview of what this issue is."

"Key Air is a charter company out of Oxford airport and they were asked, from time to time, to provide charter air service for the governor's office. And on one of the occasions they did so they did not send a bill or get paid—which was a trip to Las Vegas."

As Reich spoke from a lectern, he presented Key Air documents on a large screen and questioned Mintz about their meaning and accuracy in the bland manner of legal proceedings. (They had had virtually no time to prepare this testimony, and Reich had no obligation to alert the governor to their recent discovery before he was questioned about Key Air.)

Hearings, like trials, are hardly electric affairs. Staged, formal, and prosaic, they allow for few laughs, or even smiles. Voices are set to monotone. Postures are stiff. Meter is simple, tempo is slow, and drama is diligently withheld. In trials, explicit drama usually comes during opening and closing statements when conclusions are drawn, opinions offered, and emotion, finally, unleashed.

According to the evidence presented by Reich and Mintz, late in the evening of November 14, 2000, Rowland flew in a Hawker 800 with four other passengers—Vincent DeRosa (then Rowland's state police chauffeur), Ralph Carpinella (a friend of the governor's), and two unidentified guests—to Las Vegas with a refueling stop in Salina, Kansas.

Two days later, the plane flew to Tampa, where on November 17 there was to be a meeting of the Republican Governors

Association. On the final leg of the trip, from Tampa to Connecticut, the plane carried two passengers, according to the flight log. Three people, presumably including the governor, remained in Tampa.

Mintz's testimony galvanized the committee. Seven of the ten members asked spirited questions—about the price of the flights, who paid for them, why DeRosa was on the trip, who the other passengers were. At one point, the committee co-chair Representative Arthur O'Neill said, "Whoa, Tampa's got everybody's attention!"

After one question about whether he identified the two other passengers on the Las Vegas flight, Mintz, muffling a smile, replied, "No, uh, I guess I should say this information just has come out recently."

(Reich did present evidence that Rowland had alerted the necessary officials that he was leaving the state, as he was required to do by the state constitution.)

Mintz also interviewed Key Air's vice president Brian O'Neil, who acknowledged Key Air "had not sent a bill." O'Neil had "no explanation" for why that occurred. He could not recall any other flights for which the airline had not charged Rowland.

O'Neil, according to Mintz's testimony, fielded all the flight requests from the governor's office, and most had come from Rowland's executive assistant, Christine Corey (who, with her husband, gave the governor the hot tub on behalf of Vincent DeRosa), but some flight requests came from DeRosa, who, after working as the governor's driver until 2001, became head of his security detail and then deputy commissioner "in charge of dignitaries" in the Department of Public Safety.

(After September 11, 2001, DeRosa was appointed Connecticut's first homeland security director. He retired a few months before the committee's hearing, in March 2004, after it was disclosed that his part ownership of an auto dealership violated state police regulations.)

In the days following Mintz's testimony, his team continued to investigate the governor's relationship with Key Air. They discovered that Key Air and another company, Keystone Aviation Services Inc., shared three officers and directors (O'Neil, for example, was vice president of Key Air and president of Keystone Aviation Services) and that Keystone Aviation Services received $115,668 from the state in the 2000–2001 fiscal year and a loan guarantee of $675,000 from the CDA in 1996.

The *Hartford Courant* later revealed that the two men listed as applicants on Keystone's loan guarantee, both officers of the company, were also contributors to Rowland's campaigns.

Mintz also found evidence that the Rowland administration proposed legislation in 2002 that granted tax exemptions for "certain aircraft-related services" retroactively to 1994. Although the bill never became law, the exemption provision was enacted as part of the state's 2002 budget bill "at the request of the Administration."

In its final report, the committee concluded that "the recollection of participants in the legislative process is that the retroactive exemption specifically was designed to benefit an Oxford, Connecticut–based aircraft company. While time prevented the Committee from determining whether that company was Key Air, the Committee notes that Key Air is based in Oxford, Connecticut."

In the impeachment committee's probe, in the federal criminal investigation, and as the media's reporting expanded, evidence was presented that for years Rowland and certain people from his inner circle had been accepting gifts and bribes, often from close friends, often in contravention of state ethics rules, and often covering them up.

Among the gifts Rowland accepted were socks, a cigar, champagne, a cookie, a canoe, and gift certificates for clothing. His behavior appeared to be congenital. Reich noted at one point during committee testimony that Rowland paid $2,000 to settle a 1997 investigation into concert tickets he received.

Some of the gifts were so minor, one sensed that Rowland was either incorrigible or impoverished or both. Kathleen Mengacci, the governor's personal secretary (whose husband was named a judge by Rowland, although he was not confirmed by the state legislature), gave Rowland $3,250 "when the governor was too broke to pay his bills," according to the *Hartford Courant*. All of this was to have occurred while Rowland lived in a Georgian Revival Colonial in Hartford with a library, a sunroom, nine bathrooms, and nine fireplaces.

On the ninth day of the hearings at the capitol, the public

session lasted less than five minutes. By then, the committee had issued 140 subpoenas, conducted hundreds of interviews, gathered about 409,000 pages of documents, and deposed five witnesses. Reich's inquiry, which, in addition to Mintz's team, included about a dozen lawyers, had cost the state $6.5 million. That morning, the committee learned that Rowland's lawyers were unavailable and then convened an emergency meeting.

At 6:00 p.m. on that day, Rowland appeared on a terrace outside the governor's mansion and gave a short speech during which he resigned, effective July 1, a few days away. He added a line that was not included in the official text of the speech later released by his office: "I acknowledge that my poor judgment has brought us here."

(Rowland's lawyer Ross Garber said the resignation was prompted neither by the criminal investigation nor by a court ruling that required Rowland to testify before the impeachment committee.)

That December, Rowland declined to go to trial and pleaded guilty in federal court to one count of conspiracy to defraud the United States for accepting about $107,000 in vacations to the Vermont and Florida homes of Tomasso, two flights on Key Air (a second was to Philadelphia), and renovations to his cottage.

The plea agreement noted that among Rowland's crimes was taking no "corrective action" to prevent Ellef from rubber-stamping Tomasso's bid to build the juvenile school.

A few months later, Rowland was sentenced to a year and a

day in prison and four months of home confinement. He was assigned Bureau of Prisons Register Number 15623-014 and spent about ten months in the Loretto Federal Correctional Institution, a minimum-security prison about eighty miles east of Pittsburgh, where he swept floors, mopped, read, and exercised.

Approximately $91,000 of the $107,000 in gifts or gratuities the governor accepted and failed to pay taxes on came from Key Air. Eighty-five percent of the value of wrongdoing calculated by the U.S. Attorney's Office, which had declined to collaborate with Mintz, came from evidence he had uncovered. And it all began in a dark bar on a country road with a secret source over a tumbler of bourbon.

The Rowland committee's final report states what seems an obvious point: that Connecticut law was designed to "ensure that public servants act in the best interests of the state, and not in their own self-interest."

The committee's investigation was undertaken, in part, to ensure that the law was enforced, and Connecticut legislators were keen to demonstrate to constituents that their government provides a "level playing field for those seeking access to state benefits and services."

Reich and Mintz exposed fraud and waste by the senior member of the state's executive branch. They prevented further abuse of taxpayer funds. And, perhaps, they created a deterrent to future elected officials and government contractors looking for an illegal edge over their competitors.

(There was no deterring Rowland: in 2014, he was indicted

on unrelated federal charges connected to an election fraud involving a congresswoman. He was sentenced to thirty months in custody and released in May 2018.)

There were some structural reforms, too, that came from the committee's investigation. Since 2004, the process by which private businesses obtain Connecticut state contracts has shifted, and some agencies have been reconfigured.

In 2012, Connecticut Innovations, a quasi-public entity first formed by the state legislature in 1989 to help burgeoning technology companies, merged with the Connecticut Development Authority (whose board Rowland swayed to help his friend Robert Matthews). Connecticut Innovations now bills itself as the "leading source of financing and ongoing support," using equity, debt, and grants, "for Connecticut's innovative, growing companies." While the CDA supported distressed businesses, Connecticut Innovations focuses more on early-stage companies with creative business models. (Small businesses in Connecticut can also seek loans and other support from the Connecticut Department of Economic and Community Development.)

It is not common practice for private investigators to be appointed by independent legislative committees, and the appointment of a special counsel is also rare. (At the federal level, a special counsel can be appointed by the Attorney General's Office or by Congress, which has other, broader powers to investigate wrongdoing. State legislatures function similarly and occasionally bring in outsiders to help them probe misdeeds.)

Ten years later, Mintz and Reich worked together again

with a similar mandate from a public body. In 2013 and 2014, the Mintz Group, working under Reich's direction, was retained by a special investigative committee of the Utah House of Representatives to look into Utah's attorney general John Swallow, whose alleged corruption led to his resignation.

Private detectives, broadly defined, are deputized by lawmakers under other circumstances. Many firms that win such contracts are staffed by former prosecutors and law enforcement officers whose experience, and objectivity as third parties now that they are in the private sector, qualify them for the work.

In 2013, for instance, a federal judge in Louisiana appointed the former FBI director Louis J. Freeh, who runs the Freeh Group International Solutions and was a partner at the law firm Pepper Hamilton, to look into "alleged misconduct and monitor the claims payout program" that was put in place after the BP oil spill, according to Freeh's website. Freeh was given the title "special master" in this role.

Among Freeh's findings was that "a $357,000 payment in a damage claim arising from the 2010 BP oil spill must be repaid because the claim was fraudulent."

Of course, in the years following the election of President Trump, the national political conversation was focused on the integrity of the most senior member of the most powerful executive branch in the nation.

Robert Mueller, who succeeded Freeh as the director of the FBI in 2001, was appointed special counsel by the deputy attorney general in 2017 to investigate Russian efforts to influence the 2016 U.S. presidential election, among other matters.

Although Mueller would likely not describe himself as a private detective, he had a team of employees, mostly former prosecutors, who effectively played that role.

Mueller's work led to the convictions of many in Trump's inner circle, including the president's former campaign manager, national security adviser, and personal lawyer. Mueller achieved this, like Reich and Mintz, by ethically and painstakingly collecting evidence.

Not all fact gatherers are so scrupulous. In 2008, when a producer on a BBC documentary could not find hard proof that an English garment manufacturer was illegally employing underage laborers in India, he devised a work-around: fabricate the evidence. It took a team of private eyes on two continents to expose the truth.

9.

BANGED TO RIGHTS

On February 24, 2008, Dan McDougall, a British journalist on assignment in southern India, sent an email to his editor at *Panorama*, a BBC television program.

"I'm still in Tirupur," he wrote, "and heading to Bangalore a.m. Radio silence until now as I wanted to gather as much as I could before I contacted you—I hate giving updates when there is no update. . . . I have been painstakingly tracking down the chain of Primarks [*sic*] Indian suppliers and confident I have them banged to rights for sub-contracting not only to slums but to Sri Lankan refugees who are drifting to Tirupur to carry out what is effectively slave labour. I have filmed two separate groups of children, aged between 9 and 13, hand beading and stitching Primark Summer Season blouses, the first group are Sri Lankan refugees, living in a colony . . . about 80 kms out of Tirupur, hand-stitching childrens [*sic*] blouse vests for 2 rupees per item. . . . The Second group is in a slum on the outskirts of Tirupur, hand beading sequined [*sic*] women's tops . . . —they have never been caught with anything as bad as this."

McDougall was traveling with another reporter, Tom Heap, and they were posing as European buyers of wholesale clothing, whose subterfuge included a fake company called Rockwear Trading, a fake website, fake business cards, and a hidden camera. McDougall had just obtained proof that Primark, the British "fast fashion" retailer, was violating the U.K.'s child labor laws.

For more than two decades, *Panorama* has been the BBC's flagship investigative and current events show. According to an editorial published in *The Independent* in June 2011 by Tom Mangold, who worked for *Panorama* for thirty years and became its chief correspondent, the program was the "gold standard" of television journalism. "It has been unimpeachable: well researched, at times groundbreaking, and, above all, accurate."

McDougall, whose reporting in India was also for a companion print article, was the kind of intrepid fact gatherer *Panorama* hired to buttress its unblemished reputation. In October 2007, he published an article in *The Guardian* titled "Child Sweatshop Shame Threatens Gap's Ethical Image" exposing the "tragic consequence of the West's demand for cheap clothing."

McDougall wrote, "Despite Gap's rigorous social audit systems launched in 2004 to weed out child labour in its production processes, the system is being abused by unscrupulous subcontractors. The result is that children, in this case working in conditions close to slavery, appear to still be making some of its clothes."

In 2009, he was awarded a British Press Award for projects commissioned by *The Observer*, *News of the World*, and *The Mail on Sunday*'s *Live* magazine. He won an Amnesty International Award for Outstanding Human Rights Journalism. A posting on the World Economic Forum's website states he has reported from more than a hundred countries and war zones and "exposed trade in Zimbabwean blood diamonds" and "heroin smuggling from Afghanistan to Russia."

Not long before McDougall's report aired on *Panorama*, the BBC informed Primark it intended to reveal that, among other details, Primark was subcontracting to underage laborers in India working out of dingy sweatshops and their own homes, out of view of auditors, labor activists, and human rights groups.

The show hinged, in part, on McDougall's footage of three boys testing the sequins on brown women's vests for Primark's Atmosphere brand. (Indian law forbids children under the age of fourteen to work.) Primark asked the BBC not to broadcast the film and to turn over its evidence so Primark could either test it or respond to it. The BBC refused to disclose video excerpts, but it did provide still photographs.

Primark executives felt they could not adequately assess the BBC's claims with so little material, so before the show aired, they contacted the BBC to complain. Frank Simmonds, *Panorama*'s deputy editor, to whom McDougall reported, rejected the complaint.

*P*rimark is a chain of clothing stores formed in Dublin in 1969, when it opened its first store, and it moved to England in 1973, expanding rapidly, often through acquisitions. Today the brand has more than 350 stores in nearly a dozen countries. The United States is among its fastest-growing markets. Its brisk sales of low-cost, high-volume clothing, accessories, shoes, undergarments, children's wear, and home and beauty products compete with Forever 21, JCPenney, Marks & Spencer, H&M, and Zara. Its website states, "Adored by fashion fans and value seekers alike, Primark is widely established as the destination store for keeping up with the latest looks without breaking the bank."

In 2006, Primark joined the Ethical Trading Initiative, an international alliance of companies, unions, and NGOs that promote respect for workers' rights. In plaques posted in its stores, Primark pledged its commitment to "monitoring and progressively improving the conditions of the people who make products for Primark."

On June 22, 2008, McDougall published "The Hidden Face of Primark Fashion," in *The Observer*, a British newspaper. He set up his target: "When Primark was launched, its flagship store in London's Oxford Street was besieged by stampeding bargain-hunters and sold more than a million garments in its first 10 days. The opening drew a bigger crowd than that managed by Topshop's much-hyped launch of its Kate Moss collection, which featured the supermodel herself moodily posing in its windows. Fashion bible *Vogue* gave a Primark jacket high-end credibility."

McDougall detailed the perils of reporting by referencing another story: "I was badly beaten for being found inside a sweatshop in the lawless Haryana state border area of northern India. An angry mob chased me through the ancient alleyways, a no man's land for foreigners and police. They smashed photographic equipment and threatened to kill my translator, who had his eardrum perforated in the attack."

He presented his thesis: "Last week, in an announcement that effectively pre-empted publication by *The Observer* of this investigation, Primark announced it had sacked three of its clothing suppliers in India after being told by the BBC's *Panorama* programme of evidence that it was subcontracting labour to child workers. The investigation found that in the refugee camps of southern India young children had been working long hours in foul conditions."

Meanwhile, the BBC posted a promotion on its website promising its upcoming episode of *Panorama* would put "Primark's claims that it can deliver cheap, fast fashion without breaking ethical guidelines" to the test.

The day after *The Observer* article was published, BBC One broadcast on *Panorama* "Primark: On the Rack," which showed McDougall's evidence that Primark's supply chain included illegal labor activities. BBC executives had approved an hour-long show; most *Panorama* episodes run half that length.

The exposé was laced with sensationalistic sound bites about how *Panorama* had uncovered the "critical link between poorly paid child workers and your shopping bag in Britain" and how "carefree glamour can only come from hard, human labor." Primark, according to the show, was motivated purely by profit, and either knew and denied, or should have known and acted on, the fact that its products were tainted by child exploitation.

Primark executives knew they employed garment workers in Tirupur, a textile hub in the southern Tamil Nadu state, but they were baffled by McDougall's claim to have discovered impoverished children working for its manufacturers in Bangalore, three hundred kilometers north. The report made Primark suspicious of the show's claims.

A week after the show aired, Primark filed another complaint, this time with the BBC director general, citing inaccuracies in the BBC's reporting. When that complaint was rejected, Primark elevated its protestations to the BBC's Executive Complaints Unit. Once again, its objections were dismissed. The BBC, it seemed, was standing behind McDougall's report.

"Primark: On the Rack" had a powerful effect. The garment sector in Tirupur, already politicized because of alleged violations of labor laws, attracted even more attention from foreign media. Activists and the media in the U.K. and elsewhere were galvanized. Hundreds of workers in the region lost their jobs. India's Central Bureau of Investigation began a probe of the industry. Rumors swirled that the CBI and the Tirupur police were warning locals to keep quiet about what they knew about child labor. Anyone caught discussing the issue was subject to arrest.

Primark's top brass feared their brand's reputation was being sullied. "This was not just about Primark," said Bob Randall, the head of group security for Associated British Foods, Primark's parent, which also operates a food-processing division. "It's about a global organization. We have a lot to lose, and our reputation means everything. We had to go the extra mile to recover or prevent further reputation loss."

Like many companies seeking to mitigate risk, they hired consultants: in this case, lawyers from the English firm Herbert Smith Freehills and public relations experts from Armitage Bucks, also based in England.

Armitage Bucks coaches executives on how to appear on television and make confident presentations. Its employees train press officers, and they help clients through crises by choreographing hypotheticals the way politicians are trained to counter attacks by opponents during debates. (Armitage Bucks's chief media coach is Cheryl Armitage, a former BBC journalist.)

Primark decided to independently investigate what it suspected was *Panorama*'s (and the BBC's) faulty reporting—before the story metastasized. That assignment went to Bob Randall.

Between 1975 and 2005, Randall worked for London's Metropolitan Police, where he became a detective chief superintendent, a position that sits at the top ranks of the British police hierarchy. In his role, he advised the British government on hostage negotiation and antiterrorism tactics, and he was a member of a military and law enforcement task force that included the U.K., the United States, and Australia. For a time, he lived in Baghdad working with U.S. Special Forces to hunt terrorists and rescue hostages, and he has deployed to Afghanistan and Sierra Leone. He was a man who knew how to suss out problems.

Randall's investigation of the BBC's reporting was among his first large projects for ABF, which has annual sales of about $19 billion, operations in nearly fifty countries, and more than 100,000 employees. Most of his work involved expanding and strengthening ABF's security protocols. (Minutes after we first spoke in July 2015, he flew from London to Paris to secure a Primark store that had just been raided by gunmen who took eighteen hostages.)

One of Randall's early achievements for his first private-sector employer was to assemble a network of security and investigative personnel to protect and defend the company's reputation using a global lattice of specialized contractors. A British outfit called Diligence was in that network, and Ian Casewell led Randall's operation for Diligence.

Casewell—shaved-head bald, early forties, with energy in high relief—was well equipped for the challenge. As an employee of Diligence, he infiltrated Albanian and Turkish traffickers who smuggled women into the Netherlands, enslaved them, and sold them into prostitution. Working undercover with armed security detail and running heroin dealers as agents, he uncovered trafficking routes and safe houses where victims were held. On another assignment, he tracked down counterfeiters who flooded Europe with ineffective medicines.

Casewell, with whom I worked at the Mintz Group, learned

his craft in the early years of the twenty-first century at Europol, the European Union's law enforcement agency, which helps the EU's member states fight terrorist networks, drug and human trafficking, money laundering, organized fraud, counterfeiting, and cybercrime. At Europol, he dismantled a network of smugglers who moved Pakistanis masquerading as Afghanis seeking asylum in Europe. His probe of eastern European organized crime groups led to the arrest of a suspect who bribed Olympic figure-skating judges. And he helped choreograph simultaneous arrests of scores of alleged pedophiles in EU countries to prevent them from destroying evidence.

Casewell's assignment from Randall was broad, he told me during a series of conversations. He was tasked with identifying who sanctioned and funded "Primark: On the Rack," who reported and produced it, and whether any independent production companies were involved. Randall wanted to know if the BBC sent investigative journalists to India specifically for the show and if any local contacts were used. He also wondered if an NGO or a Primark competitor had fed the BBC counterfeit information.

Casewell's priority, however, was to find the three boys featured in the *Panorama* piece sewing the brown sequined tops in Bangalore, a crucial and effective visual that Primark suspected was not authentic.

"Let me manage your expectations here," he remembers telling Randall. "You are asking us to find three unidentified Indian boys in a country with a population of more than a billion people?"

"Yes, we're asking you to do that," replied Randall. "And you might start by figuring out where McDougall operated and retrace his footsteps."

Casewell began with the primary source: the program itself. He watched "Primark: On the Rack" closely several times. There are a number of scenes in which McDougall and Heap drive to and from sweatshops, manufacturing centers, and other locales where they meet with sources. The first oddity Casewell noticed was that their driver's face was blurred.

"Why would you pixelate your fixer?" Casewell wondered. "Maybe he knew something wasn't right and the BBC was trying to conceal his identity."

Casewell then noticed that affixed to the dashboard of one of the cars was a statuette, perhaps of a Hindu god. He also noticed that the vehicle's license plate appeared fleetingly. He froze the frame and expanded and enhanced it, which gave him the plate number.

Over the course of his career, Casewell has traveled to dozens of countries to undertake investigative work, and he understands the value of local expertise. In this case, that expertise came in the form of Nitul Shah, a vegetarian, beer-drinking former accountant from an affluent Mumbaikar clan—Gujarati industrialists reputed for their entrepreneurialism. As a young man, Shah rebuffed his father's offer to join the family's auto-parts manufacturing company in Mumbai and instead set off to become a private eye.

"My family's business had no use for my intellect," Shah

told me with a laugh when we spoke on the phone. "And I didn't care about money."

There is no regulated private investigations industry in India. Becoming an investigator there is as simple as declaring yourself one. Shah and his colleagues at CreditCheck Partners in Mumbai, which he co-founded with a friend named Mukesh Bajaj, are among the few sophisticated intelligence gatherers in the country, and they disavow the label "private investigator." They prefer "troubleshooter."

During his first few years as a troubleshooter, Shah worked for Indian insurance companies and banks. He focused more on data analytics (credit reporting, anti-money-laundering pattern recognition) and due diligence than on traditional investigative fieldwork. Eventually, his portfolio expanded to more complex investigations, which was when he was introduced to Casewell.

Shah understood the meter and tone of southern India. He brought proximity to information. He was Casewell's foyer to a foreign land.

Casewell asked Shah to watch the *Panorama* program, told him about the license plate, and sent him a brown sequined Primark top like those featured on *Panorama*. He instructed Shah to go to Bangalore, which is where the boys were reported to live.

While Shah and Casewell were tracking down leads in India, Casewell's London team compiled a dossier on McDougall. "He had an absolutely glowing CV with accolades and awards, and he ran around to challenging environments and got scoops," Casewell said.

One detail Casewell's investigators found was that McDougall's experience was as a print reporter and that "Primark: On the Rack" was among his first forays into video journalism. Casewell also learned that McDougall's wife, Navdip Dhariwal, is of Indian descent and was also a BBC correspondent.

It was now November 2008, five months after the documentary aired, and Shah wasn't finding much useful in Bangalore. He visited five or six slums and a number of schools where he showed people a photograph of the three boys. "No one recognized them," he said. After four days, Shah went to Tirupur, where he was tipped off by a source in the garment industry that foreigners, including journalists, often stayed at the Velan Hotel Greenfields on Kangayam Road.

Shah booked a room there, and for three days he chatted with guests and employees, including the manager, whom I'll call Dinesh. Bribery is commonplace in India, but Shah was under strict orders from Randall and Casewell not to bribe anyone: they wanted the operation to be above reproach; they were, after all, investigating the conduct of one of the world's most revered media companies. He told Dinesh he was hoping to learn if journalists from the BBC had stayed there in late 2007 or early 2008. Although he was not explicit about it, he did not hide the fact that he was asking for confidential information.

(Investigative work is sometimes characterized as shady, because it is often assumed that private information is obtainable only through bribery, theft, or other deception like hacking, wiretapping, and impersonation. But in many cases, all investigators do is develop relationships with key sources.)

Dinesh told Shah that in February and April 2008, McDougall and other BBC journalists had indeed stayed at the Velan. (Some of the other hotel staffers remembered them too, because they were white, rude, and carried camera equipment, Casewell told me.)

Dinesh showed Shah hotel invoices from McDougall's stay to corroborate Dinesh's story. Dinesh also told Shah that the journalists used a hotel car during one stay, and he introduced Shah to the hotel driver, whom I'll call Tanvir. Shah asked Tanvir to take him to all of the places he could remember taking McDougall. Tanvir agreed, and when Shah settled into the passenger seat, he noticed on the dashboard the same statuette Casewell had seen when he had watched the film.

Shah was gathering momentum. Tanvir drove him to many places: a town called Rakkiyapalayam, a slum on the outskirts of Tirupur called KVR Nagar, an industrial area called Laxmi Nagar, the office of an NGO called SAVE, a bus station, and Pollachi, which is about sixty-five kilometers southwest of Tirupur. In most cases, Tanvir remembered, McDougall asked him to wait in the car while he explored on foot. Tanvir pointed Shah to the buildings McDougall visited.

After exhaustive attempts to interview people all over town, Shah finally found a woman in Pollachi I'll call Navya, a housewife who earned money from freelance knitting jobs at home. She remembered "the foreigners." Shah showed Navya the brown sequined vest Casewell sent him. According to a memorandum Shah later prepared for Casewell, Navya "immediately recognized it and stated that this garment was manufactured by her." (Casewell, needless to say, was thrilled by Shah's progress.)

Navya then introduced Shah to a woman I'll call Safia, who helped Navya find work and who was employed by an Indian charity called Hand in Hand that works to eliminate child labor, promote education, and empower women.

Safia told Shah she sold three brown sequined vests to McDougall for 1,000 rupees, about $15. Her price was 300 rupees each, but McDougall told her to keep the change, according to Shah.

Tanvir also drove Shah to the Bhavanisagar refugee camp, about sixty kilometers northwest of Tirupur, where they met a man I'll call Andres, a middleman who helped find, hire, and place laborers in apparel factories. Andres remembered meeting McDougall in April 2008 and even acknowledged that he appears briefly in the documentary.

Andres had seen the documentary and told Shah that portions of it did not ring true. He said that embroidery work "is always carried out by sitting on the floor as it aids in the flow of work," according to an internal Primark report describing Shah's interview with Andres. Andres explained that the kind of stitching depicted in the film "is never carried out on table tops or by sitting on a chair." These jobs "are done only by women and girls. He has never seen any boys doing this job."

Andres also said the clothing worn by the boys and the man

identified in the film as "the boss" suggested they were "not from Tirupur or Tamil Nadu" but "from the Bangalore area, 'as the dressing is more in vogue there.'"

Piece by piece, Shah was amassing electrifying evidence to support his client's suspicion that portions of "Primark: On the Rack" were staged, but he had not found the boys, and none of his sources in Tirupur knew or recognized them. Shah and Casewell understand that investigations are not linear but idiosyncratic, and they suspected there was more to learn in Bangalore. So Shah returned to the area several times. One of his goals was to find the owner of the car McDougall used.

Shah had a source within the Indian version of the DMV who helped him trace the license plate of a silver Toyota Innova that appears in the film, and they discovered it was owned by a man I will call Ahmed. Shah located his address and interviewed him, but Ahmed said he gave the car to a colleague. However, he gave Shah the colleague's name and contact information. (I will call him Marek.) Marek told Shah that he lent his car to a friend named Khaled (also a pseudonym). This is where the daisy chain ended.

In January 2009, Shah rented a room in a small hotel in downtown Bangalore, and he arranged to meet with Marek at a coffee shop nearby. Marek, acting as Khaled's emissary, explained that his friend feared Shah was a law enforcement officer. Shah patiently explained who he was, and after careful consideration Marek persuaded Khaled to meet with Shah. The following day, Marek and Khaled came to Shah's hotel room. They spent much of the day together, and at one point

Shah opened his laptop and showed Khaled "Primark: On the Rack." Khaled had never seen the film. After seeing it, he agreed to help Shah in his investigation.

(Randall authorized Shah to compensate Khaled, but only at his standard rate and not to exceed what McDougall paid Khaled. At every stage of the investigation, Casewell and Randall ensured the integrity of evidence gathering. They wanted to make sure their findings would stand up to scrutiny in a British court.)

When they broke for lunch, Shah told me, Khaled was feeling more comfortable. He reached into his bag and handed Shah some papers, which included McDougall's BBC agenda from a previous trip to Bangalore. Khaled had found it in his car: McDougall left it behind when he returned to the U.K. This was, Casewell told me, a crucial puzzle piece: an internal BBC document that helped Casewell construct a chronology that contradicted portions of the documentary.

Shah asked Khaled to drive him to all of the places he had taken McDougall, and Khaled agreed, but only if four of his friends joined them. It was clear that Khaled did not trust Shah. Shah returned to Mumbai, but the following week he came back to Bangalore, having persuaded Khaled to meet him alone. Shah brought a security guard, which Khaled grudgingly accepted.

In an affidavit Khaled later provided to Primark, he states he first met "Mr. Dan" (McDougall) by chance on December 18, 2007, in Bangalore when Khaled worked as a driver for a car rental company called Europcar.

In Bangalore, McDougall stayed at the Monarch Luxur Hotel, which arranged for four days of Khaled's services for McDougall. Khaled says that McDougall asked him to drive through some impoverished neighborhoods because McDougall "wanted to give work to poor people" and "to get some work done on export garments." McDougall told Khaled "he did not want to give this work to big factories." McDougall also "wanted to see children working."

Khaled took McDougall on tours of more than ten slums, including DJ Halli, Modi Road, Goripalya, Bommanahalli, and Yeswanthpur. They visited some of these more than once.

On about December 19, 2007, according to Khaled's affidavit, McDougall, working alone, filmed boys in a workshop in DJ Halli that specialized in embroidering for shoes. The manager, whom I'll call Soham, was not at the workshop when McDougall filmed the children, but Soham was upset when he learned what had occurred in his absence.

During the next few days, Khaled drove McDougall to other slums in and around Bangalore, India's technology hub, but Khaled suggested that if McDougall was looking for clothing manufacturers, he should visit Tirupur, where the garment industry was well entrenched.

At some point before he left Bangalore, McDougall asked Khaled to drive him back to Soham's workshop. Khaled agreed, and he states in his affidavit that McDougall met with Soham. Khaled translated between Hindi and English for them. According to Khaled's affidavit, McDougall asked Soham if he "would make some clothing garments for him." Soham said he could not, because he produced only embroidery work for shoes.

However, Soham said that if McDougall placed a large order, he would "see what he could do." McDougall did not film on this occasion, according to Khaled's affidavit. As they were leaving, McDougall told Khaled that he intended to place an order with Soham.

That evening Khaled drove McDougall to the Bangalore airport. Khaled had proven to be a valuable fixer. Before he left, McDougall told him he would hire him when he returned to Bangalore.

*I*t was now more than a year after Khaled was introduced to McDougall, and Khaled took Shah to the home of a man I will call Aadi, who is identified in the *Panorama* documentary, Shah later learned, as "the boss." In one scene, Aadi is shown overseeing the three boys testing sequins on the brown tops. Khaled had taken McDougall to Aadi's home in January 2008, so he remembered the address. Khaled found Aadi through his contacts in the garment industry. But since then, Aadi had moved, so Khaled drove Shah to another residence he remembered visiting with McDougall. There they spoke with a few neighbors, and one of them, a woman I'll call Saira, knew Aadi's new address.

Before they went looking for Aadi, however, Saira told Shah that when she met McDougall at her home, there were several women and children knitting garments and McDougall filmed the scene. A small crowd of neighbors and children gathered outside, and McDougall "signalled to some of the children, mainly girls, to join the ladies who were working on the garments," according to Shah's report. "McDougall directed the girls to sit next to the ladies and pretend to be working on the garments; this he then filmed."

Saira then took Shah and Khaled to Aadi's home. According to an internal Primark memorandum, to secure Aadi's cooperation, Shah adopted a cover story "without disclosing any sensitive information and preserving the Client's confidentiality." As with many of the sources Shah tracked down, Aadi was at first reluctant to speak, but he eventually agreed. He said he met McDougall only once, in January 2008.

McDougall introduced himself to Aadi as a "garment buyer from Europe" and said he "represented a very large buyer who was regularly sourcing garments from one of the local garment suppliers in Bangalore." McDougall hoped to build a direct relationship with Aadi and to eliminate any middlemen. He offered to triple what Aadi's local buyers paid. According to the Primark memorandum, McDougall told Aadi this money would help him "and others in his position to improve their social and financial status." Aadi had no reason to doubt McDougall's intentions and agreed to cooperate. Aadi's wife then drove McDougall to a few manufacturing sites, including Soham's workshop.

Khaled told Shah that when McDougall visited Bangalore in February 2008, they returned to two of the slums they visited the previous December, DJ Halli and Bommanahalli.

Khaled took Shah to DJ (Devara Jeevanahalli) Halli and to a building he remembered having visited with McDougall. Many sections of DJ Halli—where most residents are from a lower caste called Dalit, which means "oppressed"—are made up of warrens of crisscrossing lanes too narrow for cars. They are cramped with squat, shabby homes. Garbage is strewn alongside buildings and piles up on streets and is picked through by goats and chickens.

When Shah and Khaled arrived at the building, Khaled asked Shah to wait in the car while Khaled went inside. After a time, Khaled emerged with Soham, the sweatshop owner. At first, Soham was frightened. He feared Shah was a labor inspector. But Shah convinced him that wasn't the case in part by showing Soham excerpts of the documentary on his laptop.

Soham recognized the boys and McDougall in the film, and he gave Shah a tour of his workshop. He told Shah it was the one that appears in "Primark: On the Rack." Shah was struck by its sparseness. There were no hordes of children packed into a warehouse, no brown vests, no Primark clothing. There were only five or six boys. Then Soham introduced Shah to the three boys who appear in the film. Shah compared them with screenshots from the documentary.

"When I saw them, I felt privileged," Shah told me. "It was a magical moment. I believe in God, and I had no idea how to express this feeling."

The boys were nervous, so he gave them some chocolates. Shah showed the boys the film, which they had never seen, and spent a few hours with them. He took photographs of them, so he could prove to Randall and Casewell that he had found the three boys he had been looking for among the hundreds of millions in India. The investigation had been going on for more than a year at this point.

Soham explained that his workshop specialized in embroidery work for women's shoes and that his employees did not manufacture garments and were not even capable of such work. He confirmed that McDougall visited his shop on February 25, 2008, spoke with his employees, including the three boys, recorded video of them at work, and then left.

Soham told Shah that McDougall gave the boys brown sequined vests—the same vests McDougall bought from Safia the day before in Pollachi, near Tirupur, three hundred kilometers south—"and asked them to pretend to work on [them] while he filmed." In his affidavit, Khaled stated he "translated this into Hindi for the boys and said something like 'do this action' and made an up and down action with my hand which the boys copied."

Khaled also confessed that he helped McDougall further construct his ruse. At McDougall's request, Khaled instructed the boys to "keep quiet" and "continue working," which can be heard in the film and seen in English subtitles. Khaled said the whole video was "staged."

Shah wrote up his findings and sent them to Casewell and Randall, who flew to India to confirm what Shah uncovered and to meet many of the sources Shah had tracked down, including Soham and the boys. Later, Shah flew to London to brief Randall and Primark's lawyers on what he'd learned.

As a result, Primark submitted to the BBC's Executive Complaints Unit, or ECU, "a substantial quantity of fresh evidence that gave rise to further grave concern over the authenticity of the footage used in the programme," according to a Primark statement. In July 2009, the ECU opened an inquiry.

The ECU investigated, and although it "admitted there were serious shortcomings and unanswered questions about the way the programme was made," it rejected the complaint. Primark appealed to the BBC Trust, the governing body tasked with making sure the media organization delivers on its mission to "inform, educate and entertain." The trust agreed to consider the evidence.

As part of what evolved into a joint probe (the BBC also investigated Primark's complaints), the two sides shared information. As a result, Casewell's and Primark's lawyers (from Herbert Smith Freehills) received a tranche of emails and videos, among other documents.

When filming, McDougall normally used a different tape each day, which was standard practice. But when Casewell and the lawyers analyzed the metadata, or digital date code, from the tape that was purportedly used to shoot footage on February 24, 2008, in Pollachi, near Tirupur, they discovered it was stamped February 25—the day McDougall filmed the three boys in Soham's Bangalore workshop, which is three hundred kilometers away.

McDougall had reused a tape, Casewell concluded. The effect was "to make it look like continuous footage," according to Primark. The email McDougall sent to *Panorama* on February

24, said Casewell, was clearly designed to fool his editor; Mc-Dougall planned to stage footage. (McDougall's defense was that he made a mistake in that email.)

Shah arranged for Casewell to meet with the three boys in a hotel conference room in Bangalore. Casewell set up a video camera and recorded the interview with them. (He spent hours preparing with Primark's lawyers to ensure that any questions he asked were not leading and so would be admissible as evidence in a British legal proceeding.)

In the interview, one boy identifies McDougall from a photograph Casewell gives him. Casewell also hands the boy a Primark vest, and the boy says McDougall told him to pretend to work on it. Another of the boys tells Casewell that before he met McDougall, he had never worked on such a garment. All three boys said they specialized in women's footwear—not sequined tops.

The video also shows an isolated frame of the boys "testing" the brown sequins. Underneath the sequined garment and one of the boy's arms, there is a cloth that is sold to Indian retailers for *chappal*, or footwear, not sequined clothing, according to Casewell's video.

(Randall, who died many years later, in 2016, separately hired an expert in garment manufacturing in India who provided a professional opinion stating that none of the boys was performing legitimate stitching techniques.)

In June 2011, the Editorial Standards Committee of the BBC Trust issued a final report. While *Panorama* obtained "clear evidence that work was being outsourced from factories in India in contravention of Primark's own ethical trading principles," the BBC found "serious editorial failings" in Mc-Dougall's piece. The report stated, "It was more likely than not that the Bangalore footage was not authentic." The BBC was convinced that "the distance between Pollachi, where [Mc-Dougall] had already filmed women working on the brown sequined vest tops, and Bangalore made it improbable that he had found the same tops in these two locations on successive days."

The BBC conceded that "no brown vest tops other than the ones being worked on by the boys can be seen in the Bangalore footage, whereas it is likely that, if a quality control process was being undertaken, the workshop would have been handling a significant number of the garments."

It also acknowledged that the short sequence filmed at a tight focus was inconsistent with other footage McDougall took.

The BBC agreed with Primark's contention that the adult heard speaking to the three boys in a local language or dialect subtitled as "get on with the work, little boy," and "keep quiet and get on with the job, boy," were spoken not by the boys' supervisor, as was presented in the program, but by Khaled. This "constituted a breach of the accuracy editorial guideline," the report read.

As a result, the BBC's Editorial Standards Committee, or ESC, vowed that the BBC would review how it handled complaints and evaluate its journalistic standards and editorial policies. The BBC said it would ensure "staff involved in investigative reporting . . . understand their responsibilities when it comes to authenticating evidence."

Helen Boaden, chief of BBC News, said the BBC and *Panorama* brands were "strong and robust" because "we know that when we make a mistake, we own up to it and we learn from it." The BBC ran an official on-air apology on its flagship daily news show at 6:00 p.m. on BBC One, which stated, in part, that the faked footage "caused unfairness to Primark" and "should not have been broadcast. The [ESC] considered that there was a strong public interest in *Panorama* conducting such investigations. However, there were serious failings in the making of this edition of the programme and they are not acceptable. The BBC Trust would like to apologise unreservedly to Primark and to our audiences."

According to Shah and Casewell, when McDougall heard his conduct was under investigation, he contacted Khaled (who at the time had moved to Saudi Arabia but soon returned to India in search of employment) and asked him to lie about what he had done with McDougall and what he had witnessed.

With no apparent sense of irony, McDougall accused Shah of fabricating evidence. McDougall and his *Panorama* colleagues refused to be interviewed by the BBC Trust. Instead, they provided written responses "with the benefit of legal advice," the BBC disclosed.

(*The Observer* article McDougall had written that had been published the day before "Primark: On the Rack" aired also included information that could not be verified, according to a Primark document, which stated, "The article features a girl called Mantheesh, said to have been in the Bhavanisagar refugee camp but who is not on film. Dan McDougall declined to provide any notes or evidence to the BBC Trust to support the content of this interview with this girl.")

McDougall issued a public statement, saying that he was "appalled" by the BBC's decision. "I have rarely seen a finding so unjust in outcome, flawed in process, and deeply damaging to independent investigative journalism." As of this writing, he covers Africa for the London *Sunday Times*.

*H*ad the phrase "fake news" been employed in the cultural lexicon with the ferocity it has been in recent years, Primark and McDougall could have weaponized it against each other.

Primark hired Casewell's team to help it protect the integrity of its brand. Although the company has faced other controversies—it was one of many retailers that sourced garments from the Rana Plaza factory in Bangladesh that collapsed in 2013, killing more than eleven hundred workers—in this case, the company was sufficiently galvanized that it funded a global investigation to counter a report by one of the powerhouses of British journalism.

Private investigators are a crucial component of such brand protection. Among our clients in this realm are crisis management and communications companies, a relative of traditional public relations firms that are often the recipients of the first phone call when a large company or wealthy person is facing controversy.

FleishmanHillard, Ketchum, BerlinRosen, Portland Communications, Sitrick, and LEVICK are just a few such firms in this expanding field. They employ former journalists, government officials, policy advisers, data scientists, public relations specialists, lawyers, and others to manage threats and frame potentially damaging news that could affect the lives and livelihoods of individuals and corporations.

FleishmanHillard, for instance, provides advice on how to handle workplace accidents, disputes with labor, litigation, management shuffles, government probes, product failures and recalls, natural disasters, workforce reductions, and activism, according to its marketing materials. LEVICK's "24/7 crisis response team" gives clients "full-scale tactical and strategic support" for communicating during a crisis including "24/7/365 rapid-response crisis management" teams, response training, message development, media relations, social media management, online monitoring, third-party ally development, a crisis response "War Room," and online reputation management.

Some communications firms have suffered when they were unmasked as propagandists for clients or found to have used unsavory tactics to intimidate, for instance.

In 2018, *The New York Times* labeled Bell Pottinger the "P.R. firm for despots and rogues" after it "helped drive racial tensions" in South Africa on behalf of corrupt government contractors "to levels not felt since apartheid." Bell Pottinger, once considered the U.K.'s most respected public relations firm, folded under the media's exposure of its clients and tactics.

Primark used the crisis communications firm Armitage Bucks, in part, to produce Casewell's video to rebut *Panorama*'s version of events. It was posted on a website that included a Primark statement and a timeline, and it can still be found on YouTube.

Armitage Bucks's strategy was to flood the public domain with evidence exposing how Primark was victimized and what it did to counter the attack on its reputation. The goal of such a media strategy is to ensure not only that threats are addressed in real time (often directly to the public, instead of through the filter of a journalist) but also that any misinformation is parried into oblivion.

McDougall, too, vehemently stood by his reporting. It is conceivable that he, too, has drawn on consultants to support his position in the public eye: his Wikipedia page is conspicuously full of quotations from his defenders.

The malleability of the definition of "fake news" has severe consequences. The term seems to ring clear enough: factually incorrect reporting or propaganda. But it has been stretched and manipulated by everyone from reporters to consultants to politicians. (President Trump—taking a page from the despot's handbook—regularly, and intentionally, garbles "fake news" to attack accurate media reports that are critical of him. And he has demonized the traditional media to great effect using nothing more than the sheer force of his personality.)

Some attempts are being made to address this morass. In October 2018, the British government banned the term "fake news" from appearing in official documents, instructing ministers to instead use "misinformation" or "disinformation." (According to a statement that accompanied the ban, "fake news" is "a poorly-defined and misleading term that conflates a variety of false information, from genuine error through to foreign interference in democratic processes.")

Although established media brands like *The New York Times*, *The Washington Post*, and *The Guardian* still have robust teams of investigative journalists and nonprofit journalism outfits (such as the Center for Investigative Reporting, *ProPublica*, and the Marshall Project) that not only have proliferated but often collaborate with traditional media companies, the industry at the local level has been decimated.

There is no reliable data on how prevalent McDougall's brand of fakery is—that is, making up information and passing it on to readers or viewers as real. Past famous examples include Janet Cooke of *The Washington Post*, who won a Pulitzer Prize in 1981 for a story she made up about an eight-year-old heroin addict and doctored her own résumé; Stephen Glass, who published a number of articles in the 1990s in *The New Republic* that were later proven to have included fabricated quotations, sources, and events; and the photojournalist Eduardo Martins, who, until he was caught in 2017, tricked photo news agencies and media companies into believing he was a conflict photographer (he had actually just stolen and repackaged pictures taken by other photographers). But even one such story is detrimental to the functioning of a democratic society and a free and accurate press.

Casewell's investigation reminds us that the news we consume is not always trustworthy and that private investigators occasionally help to restore that trust. If the public does not have faith in the media and if citizens in a democracy do not have a free press, they are not informed citizens. Uninformed citizens cannot make wise choices about how society is governed, which can lead to anarchy or dictatorship or both.

*N*efarious behavior is not confined to the U.K. and India, of course. Without the doggedness of a young labor union investigator in New York, a brash executive's tabloid-ready crimes might have gone unchecked.

10.

MARCH ON THE BOSS

In the spring of 2002, an obscure military contractor
called Point Blank Body Armor in Pompano Beach, Florida,
began to attract national media attention. *Time* magazine reported that American soldiers in Afghanistan were praising the
Interceptor, a light flak jacket manufactured by the company
that contained Kevlar, ceramic plates, and pouches for muni-
tions, handcuffs, two-way radios, and gas masks. The jacket
was credited with protecting soldiers from grenade explosions,
artillery shellings, and bullets. The *Associated Press* published a
profile of Point Blank, and it was featured on the *CBS Evening News*.

Shareholders of Point Blank's parent, DHB Industries Inc.,
were rewarded by the attention: revenues increased 40 percent in
2001, and within six months of the September 11 attacks, shares
of DHB's public stock increased in value by more than 200 percent. Terrorism had galvanized America's warfare industry.

DHB was formed as DHB Capital Group Inc. in Old Westbury, New York, on Long Island in 1992 by its namesake David H. Brooks, a former securities broker from Brooklyn. Brooks's strategy for DHB was to acquire insolvent companies in the defense and apparel industries, assume their debts, and secure government contracts for the companies' products. Point Blank was acquired out of bankruptcy for $2 million in 1995. Among other subsidiaries, DHB owned an athletic-equipment manufacturer, and it sold various kinds of protective gear to SWAT teams and police forces.

Brooks was born in 1954 and raised by working-class Jewish parents under traumatic conditions. His mother, Anna, survived four years at Auschwitz during World War II, where she was sexually abused, according to court records. Near the end of the war, she escaped, eventually settling in New York, where she married Joseph Brooks. Joseph, who earned a living as a cabinet-maker, suffered from a genetic bone disease that stunted his growth—and unwittingly exposed his son to constant ridicule.

Friends taunted Brooks about his father's height (he was only four and a half feet tall), calling his dad a "midget" and a "freak." Anna, who was emotionally distant and suffered from anxiety and paranoia, was unfaithful to Joseph, with whom she often fought. Brooks's older brother inherited Joseph's brittle bone disease and later tried to commit suicide.

"As a child, people called my family 'diseased,'" Brooks once told a psychiatrist, according to statements in a publicly available court record.

But Brooks was industrious. During elementary school, to make money he sold fireworks, mowed lawns, painted houses, washed cars, built docks, and shoveled manure from horse stables.

Reflecting on his teenage years, Brooks confided to a therapist, "I just started realizing really young that I had to know numbers and money a little bit because maybe money could get me out of the situation that I felt I was in. . . . It could make my dad six feet or a foot taller than he was. . . . It would give me something that would excuse the other parts of what I had, of where I came from, what I was, or the environment. So money became, like, sacred, and I tried to make it in every single way I could. And that became more important to me than anything else."

He found solace in legal and illegal gambling: in high school he learned to trade stocks and spent much of his free time at the racetrack.

As he grew older, Brooks developed a dysfunctional relationship with money. He told one therapist he "couldn't stand monthly bills." He often paid cash for what he bought. He once paid $196,000 in cash for a Rolls-Royce Corniche. In 2002, when his income soared as a result of DHB's success, he bought his family a Learjet 60 airplane for $7 million in cash. In 2006, he and his brother, Jeffrey, rented the penthouse in the Bloomberg building in midtown Manhattan for $75,000 per month. The same year, his family paid $125,000 per month to rent a twelve-thousand-square-foot town house in London not far from Hyde Park and Buckingham Palace. The home came with a thirty-five-foot swimming pool in the basement.

Brooks shopped in bulk. He would buy six bottles of cologne, thirty tubes of skin cream, forty pairs of underwear, a dozen pairs of sneakers. He would buy every shirt in his size that a store stocked. He would buy ten ink refills for every pen he owned. To his therapist, he estimated that he owned between five and ten million pens. He would attend ten Madonna concerts in a single tour and buy shopping bags full of merchandise. "I'll take everything you got in XL," he would say to a clothing store employee.

Shortly before Brooks got married, his mother revealed to him that Joseph, the man he grew up calling "Dad," was not, in fact, his biological father. Anna confessed that she conceived Brooks and his younger brother through artificial insemination; she didn't want them to inherit Joseph's brittle bone disease.

If anything, the revelation drew Brooks closer to Joseph. In September 1992, as Joseph lay on his deathbed, Brooks got down on his knees. "Two days before he was dead," Brooks told Joseph that "I was going to turn shit into gold and build a billion dollar company for him . . . in his honor."

Three weeks later, Brooks filed incorporation papers for DHB Capital Group.

*I*n 2002, just as Brooks was beginning to prosper, a twenty-three-year-old recent college graduate named Luke Brindle-Khym accepted a job offer as a research analyst in the strategic affairs department of UNITE, the Union of Needletrades, Industrial, and Textile Employees.

After studying English and history at Carnegie Mellon in Pittsburgh, where he grew up the eldest of four boys with fierce, working-class sympathies and an evangelical commitment to social equality, Brindle-Khym lived in Paris for a year. Inspired by George Orwell's memoir *Down and Out in Paris and London*, he spent evenings in the basement of a bistro washing dishes and drinking cheap red wine out of plastic bottles.

Brindle-Khym's only professional experience was in Lesotho, a small, mountainous nation tucked within South Africa, where he spent a few months inspecting garment factories. Stunned by the horrific working conditions he witnessed, he pledged to embark on a career in the labor movement back in the United States.

UNITE, which was formed by the merger of two of America's oldest unions, the International Ladies' Garment Workers' Union and the Amalgamated Clothing and Textile Workers Union, was a minor player in a faltering industry. Membership was in decline. Manufacturing had been shifting overseas for decades. But UNITE had a scrappy, militant reputation. In 2002, it reported having a few hundred thousand members, but, Brindle-Khym told me, it spent more money organizing than did much larger unions. "We punched way above our weight class," he said.

On his first day on the job in New York City, Brindle-Khym, who is reed thin with a frenetic energy, attended a press conference at City Hall with his new colleagues, including the union's boss, Bruce Raynor, and its southern regional organizing director, Scott Cooper. The Democratic city council member Eric Gioia, chairman of the city's Oversight and Investigations Committee who had mayoral ambitions, was at the lectern with a DHB line employee, and they were advocating that the city cancel a contract with DHB, which supplied the NYPD with body armor, because one of its vests failed a live-fire ballistics test.

"During the shooting of one of the vests," a state safety report reads, "an actual penetration occurred. The bullet went right through the protective panels. Had this been on the body of a Police Officer, this officer would have been either seriously injured or dead." Similar reports came in from the Pentagon, raising questions about DHB's military-grade vests.

It was a chaotic event attended by news photographers and elected officials. Raynor and Cooper, who had been working to organize some of DHB's workforce in Florida, had been sued by DHB for defamation. Litigation against a union by a company whose employees it targets can benefit the labor campaign: companies rack up costs as management becomes distracted.

Most companies would not have incurred such legal expenses, Brindle-Khym told me. He was learning his first lesson about corporate America: not all corporate executives act rationally.

"Brooks owned the majority of DHB's stock, but he didn't seem to run it like a public company," said Brindle-Khym. "It was his fiefdom. He took our critiques personally."

The publicity around the defamation case and the calls to end the NYPD contract helped UNITE's strategy to pressure

DHB. The company's workers had been complaining to the union for months that their managers had been pushing them to speed up production and fill orders faster, even though they knew quality control was suffering. "We thought the 'quality critique' of the vests, as we called it, was going to win the campaign," said Brindle-Khym. "DHB was churning out bullet-proof vests that didn't work and now the public knew. This was going to transform our campaign from a parochial dispute into a national story about our troops."

UNITE's strategy to gain leverage through DHB's customers was part of a "corporate campaign" in which unions map out and target the structure and vulnerabilities of companies whose workers they seek to organize and whose managers, shareholders, and vendors they seek to influence. This includes identifying relationships with suppliers, regulators, shareholders, and lenders; discerning lines of business and revenue streams; and figuring out what a company's growth plan is.

"At the time, I didn't know much about business or what it meant to be a public company," Brindle-Khym told me. "I had just spent a year living in France. I was an idealist. But this was concrete."

He cloistered and immersed himself in the job. He and Ahmer Qadeer, UNITE's director of strategic affairs, who began to mentor Brindle-Khym in the art of investigating a public company, spent their days buried in paper and on the phone, all with the goal of trying to understand how DHB operated and how it made money.

The flow of capital, disclosure requirements, debt offerings, communications with shareholders: Brindle-Khym was learning how public companies were run and transact, how money flows through them, how they take on debt, how they tell, or don't tell, their stockholders what they are doing.

He educated himself on defense contracts, federal regulation, the National Labor Relations Board, congressional budgeting, procurement standards, lobbying activity. He obtained copies of federal contracts, pored over purchase orders, perused deeds, found internet domain registration records, filed FOIA requests, and read postings in investor chat rooms.

As they had done with the police union in New York, Brindle-Khym and Qadeer also lobbied parties with more leverage than UNITE to act on its behalf. "We tried to insert ourselves into the conversation between the company and its customers," said Brindle-Khym. They contacted the Pentagon, spoke with other police unions, wrote press releases, and met with reporters.

Through contacts in the garment industry, UNITE tracked down the purchasing officers at police departments that bought DHB armor and tried to persuade them that officers might be at risk because DHB's equipment was ineffective. To make their argument, they cited the failed NYPD ballistics test results. A few departments agreed to buy armor from other vendors.

UNITE's success in lobbying law enforcement agencies was one tactic in a broad strategy, explained Brindle-Khym.

"There are three components to such a tactic: winning, being right, and looking good," he said. In the case of DHB, "we got all three."

Before Brindle-Khym joined UNITE, the union had tried to organize about three hundred workers at DHB's Point Blank manufacturing plant near Fort Lauderdale. Most of the workers were from Latin America and Haiti. The union sent undercover organizers to meet with employees.

"That factory had a reputation for being a sweatshop," said Brindle-Khym. It was, in labor parlance, a "hot shop" where workers are desperate to sign up for the union because they suffer countless indignities: low wages, few benefits, little or no time off to visit sick children, filthy bathrooms, poor air-conditioning. In fact, it was the workers who had approached UNITE seeking help.

UNITE also used what unions call a "blitz" tactic that held an element of surprise. Scott Cooper, UNITE's organizing leader, dispatched about twenty-five professional organizers and volunteer union members to South Florida. To take the initiative, Cooper scheduled the blitz to kick off at 4:00 p.m. on a Friday, just as the workers were clocking out for the weekend. The blitz organizers, most of whom spoke Spanish or Haitian Creole, fanned out across Greater Fort Lauderdale to visit workers at their homes on Friday evening, all day Saturday, and Sunday afternoon. This gave them time to meet with as many workers as possible over the weekend, before the company was able to launch its anti-union campaign.

Cooper, the lanky son of a Methodist minister who speaks with a southern drawl, recalled that first weekend. "They were ready for the union. You could see it in their eyes." The conversations in those initial house visits "were about much deeper things than bread-and-butter issues. Instead of 'I need a twenty-five-cent raise or cheaper health insurance co-pays,' workers were saying, 'My boss talks to me like I'm an animal.'" It was easier to mobilize workers "when people are talking about their families, their own dignity, their basic psychological and spiritual needs," Cooper said. "But it was more that they wanted to know 'What's the battle plan?'"

Organizers asked workers to sign "authorization cards" pledging an interest in voting to unionize. Securing authorization cards from 30 percent of workers is the minimum needed to start the election process administered by the National Labor Relations Board. But UNITE signed up 70 percent.

Meanwhile, the union organized a "march on the boss." During its morning break on Monday, dozens of employees gathered quickly, like a flash mob, and walked into the office of Sandra Hatfield, DHB's chief operating officer, to present a petition asking for better treatment. Workers later told the NLRB they were loudly but peacefully chanting "Sí se puede" (Yes we can) and other slogans.

Irked by its employees' activism, DHB went on the offensive. The company called the police and reported a "riot," and dozens of armed personnel raided the Fort Lauderdale factory. The UNITE team was taken aback by DHB's ruthless tactics.

"If you are in the C suite of a large company, you make a cost-benefit analysis" when presented with an employee demand to unionize, said Brindle-Khym. "The rational thing may be to settle with the union. You pay a bit more in wages, but have less staff turnover and avoid the legal fees and distractions. You can focus on making money for shareholders." UNITE, after all, was not asking for much: a contract that guaranteed slightly higher wages.

In response, DHB increased the pressure by firing the employees it suspected were union sympathizers. UNITE filed a complaint with the NLRB, alleging that the firings were illegal. The tit-for-tat continued unabated. Workers voted to go on strike. DHB ultimately spent more than $1 million to pay a phalanx of off-duty Broward County police officers, who wore uniforms and brought squad cars to patrol the picket lines and intimidate the strikers. The company had, in essence, hired a private security force.

DHB's calculation was rational at first glance. "Federal sanctions for employers who violate labor laws are minimal," said Brindle-Khym, who speculates that DHB was willing to pay fines (to anyone who had the power to sanction DHB) to get its way and avoid a deal with a union.

The NLRB charges took months to resolve. Even with what is called "fast track" processing, it took almost six months for UNITE's fired members to win a court order for DHB to reinstate them with back pay. DHB also continued to sue union officials, including Cooper and Raynor from UNITE, and even the head of a police union.

The adversaries were at loggerheads, especially since DHB was making statements to shareholders claiming victory over the union. Yet Brindle-Khym and Qadeer were not deterred. After a few weeks, they had focused on DHB's Point Blank Body Armor division, which had secured an exclusive

Pentagon contract to manufacture the Interceptor vest. It was by far the most profitable segment of the business: DHB's cash cow. This, they decided, would be where they would apply pressure.

Brindle-Khym also began to pursue a curiosity he'd noticed: Brooks's enthusiasm for harness racing. Brooks grew up at the racetrack. Gambling on horses was one of the ways he had made money as a teenager. Over the years, Brooks had acquired horse stables including, in the late 1990s, one in Fort Lauderdale called Perfect World Enterprises, among the largest in the world. He and his family eventually came to own a thousand horses. His stable, Brooks once said, broke every record and, for a time, dominated harness racing in North America.

During a psychological assessment approved by Brooks and his lawyers, Brooks (who, the assessor noted, often pulled on his chest hair and chewed on pens) acknowledged some "obsessiveness in horse racing." Brooks once said, "I don't think there's anyone smarter or sharper [than I am] at trading or numbers . . . and I believe it."

Qadeer urged Brindle-Khym to find people involved in harness racing who might have a motive to talk to him about Brooks: jockeys who were unionized, contractors who maintained and renovated Brooks's stables, disgruntled former track employees.

As they dug deeper into the stables' details, the two investigators marveled at the names of some of Brooks's horses: At Point Blank, D Interceptor, Precious Joe, Reimburse Me, Blood

Diamond, Master of Wars, David the Great, I Am the Prophet. And Brindle-Khym's favorites: Abolish Unions and Fire Everybody.

Brindle-Khym compiled financials on Brooks's harness-racing interests, looked for past labor disputes, identified gambling regulators that might have files on Brooks, dug up lawsuits. He acknowledges that at the time he did not know anything about the industry, so he asked gaming regulators in several states, including New Jersey, New York, and Florida, for documents that named Brooks and his companies.

One day, Brindle-Khym received a package from the New Jersey Racing Commission that contained copies of seemingly trivial correspondence it had with Brooks, including a letter describing a dispute over a licensing fee and a copy of a check for $5,207.36 made out to Freehold Park Racing LP from a company called Tactical Armor Products, or TAP.

Brindle-Khym recognized the company name, though he could not place its purpose or significance. "There were dozens of companies using Brooks's home on Long Island as their mailing address," Brindle-Khym recalled. "We didn't know how to connect all the dots."

After further database searching, he learned that TAP was incorporated in Tennessee, and documents he obtained from the secretary of state's office there (the agency in each state that maintains corporate records) showed TAP was set up in April 2000.

Other information on the documents intrigued Brindle-Khym. Jeffrey Brooks, Brooks's younger brother, was listed as the director, and Anna Jacobs, Brooks's mother, was the incorporator of TAP. Brooks's wife's name later showed up on Tennessee records as TAP's "registered agent."

Brindle-Khym also realized that another address on the TAP records matched the address of a DHB subsidiary in Tennessee: it was the office of PACA, manufacturer of the Interceptor vest.

He wondered: Why was a company controlled by Brooks's brother paying licensing fees to the New Jersey Racing Commission? What was TAP? And what was its connection to DHB? Brindle-Khym began to sketch a relationship graphic that eventually came to resemble a spaghetti bowl. It went up on Qadeer's office wall.

The TAP check Brindle-Khym discovered led him to reexamine documents—news reports, quarterly financials, corporate filings—he had previously gathered with an eye toward uncovering clues about Brooks's financials that he had missed. This is a savvy and effective investigative technique.

I have known Brindle-Khym for more than fifteen years and have met few professional detectives or fact finders—not investigative journalists, not criminal prosecutors, not game theorists, not police detectives, not spies—with his instincts. His ability to find and burrow into massive or little-known data sets and hunt through them like a bomb-sniffing dog and retrieve answers is astounding.

"Luke is relentless, has high standards, a solid moral compass," Qadeer said of his former supervisee. "And yet he remains aware of limits, both his own and what is knowable. It provides him with a rare combination of conviction, mastery, and humility when he approaches a problem."

Among Brindle-Khym's strategies (he later got a law degree) is to persuade people, deploying his close read of the law, to release information. I once witnessed Brindle-Khym chastise a county clerk in New York who tried to withhold archived court documents from him. Brindle-Khym had read the statute that states the record was releasable to a third party, and he quoted the relevant subsection and paragraph from the statute he'd found online.

Like Indiana Jones declaring that a recovered antiquity belongs in a museum, Brindle-Khym passionately explained that the documents he sought were part of the public record and must be turned over. They were.

He is also a fearless interviewer. I once watched him confront a witness during a murder investigation in the witness's front yard. Brindle-Khym's presence was not met with appreciation. As the witness was cutting his lawn with a hand mower, he jabbed it at Brindle-Khym's feet. Brindle-Khym danced around the blades while he kept up his questions.

One day Brindle-Khym rewatched a video that had been posted on DHB's website, a television news report showing factory workers in Tennessee making protective armor for American troops. The clip cast the company as patriotic and employing a proud, diverse, and diligent workforce. But with his newfound awareness of TAP, Brindle-Khym noticed that on one side of the factory floor were DHB employees sewing vests while on the other side of the factory floor were TAP employees making ceramic plates—all under one roof.

Brindle-Khym then scoured DHB's regulatory filings with an eye toward finding its disclosure of TAP's relationship to the company. He found no official connection. So the TAP discovery was revelatory.

"Brooks was buying armor from himself and not telling shareholders," Brindle-Khym said.

The seemingly innocuous $5,207.36 TAP check proved valuable in other ways. Brindle-Khym had spent so much time barnstorming through Brooks and his family members' financial forms that he began to recognize their handwriting and signatures. The check was signed in the name of Terry Brooks, Brooks's wife, but it was not signed by her, Brindle-Khym suspected; it was signed by Dawn Schlegel, DHB's CFO, who had distinctive penmanship, with large, loopy letters. (Brindle-Khym recognized Schlegel's signature because as CFO she was responsible for signing DHB's financials.)

Again he wondered: "Why was the CFO of DHB, a public company that made body armor, signing checks on behalf of Brooks's wife (on a checking account owned by a Tennessee company incorporated by Brooks's mother) to a horse-racing regulator in New Jersey?" It seemed a blithe disregard for corporate formalities: at the least a disregard for the law; at worst potential financial misconduct.

This was not the first time that UNITE's team came across odd transactions involving members of the Brooks family.

DHB rented a factory in Florida from a partnership controlled by Brooks's wife and owned by his children. The company told investors about this setup, claiming that "the terms of the lease are no less favorable to the Company than terms that would have been obtained at the time of the lease from an unrelated third party." Translation: even though Brooks has an incentive to charge DHB a higher-than-market rent (because that meant more money for his family), everything was legal.

However, when Qadeer, an economist by training, investigated, he found the opposite to be true. He surveyed real estate brokers in Broward County, Florida, and calculated the going rent per square foot for similar industrial buildings. It turned out that DHB's landlord was making about $9.40 per square foot, while most factory space in the region rented for between $4.23 and $5.40 per square foot. DHB was spending twice as much to rent the factory from Brooks's kids as it would have paid an independent landlord.

Brindle-Khym and his colleagues calculated that through TAP (a private company) and other self-dealing, Brooks was able to secretly extract millions of dollars from DHB, the publicly traded affiliate of TAP. Brooks's failure to disclose the relationship confused Brindle-Khym, in part because the company did disclose some related-party transactions, including the factory rental and an airplane Brooks rented to the company. The TAP fraud, explained Brindle-Khym, was an order of magnitude larger than published related-party transactions, including the fact that Point Blank rented factory space from Brooks's wife at an inflated rate.

"To the public, it looked like TAP was simply a vendor, but we suspected it was much more: classic vendor fraud, a hidden interest in a company, undisclosed profits." In short, Brooks was effectively paying himself through the back door: DHB to TAP.

In June 2003, UNITE sent its evidence to the Securities and Exchange Commission that DHB had violated federal securities laws. (Members of the public can submit complaints to the regulator if they believe they have evidence of wrongdoing.)

The document UNITE submitted was the third formal complaint by the union, which also issued a press release on its findings. The heart of this complaint was Brindle-Khym's discoveries, but union officials were unsure of how convincing the complaint would be.

"I felt it was powerful research," he remembers. "But we had an internal debate about what it meant and how it could be used."

Brindle-Khym called every reporter he could find who had written about DHB, its affiliates, or the labor dispute to tell them what had been filed. But his alerts were met with skepticism. "It's just a union press release," he was told.

After UNITE filed the complaint, Brindle-Khym and his colleagues impatiently monitored DHB and its subsidiaries through Yahoo Finance chat boards, the media, and the SEC. Now that they had seeded the public record with sharp, deeply sourced critiques of the company, they expected shareholders, regulators, and journalists to dig into the leads themselves. But there was no such activity.

Luke did, however, receive an email from a former DHB ballistics engineer who claimed to have damaging information. The source said he worked directly for Brooks and had salacious details to share but refused to email anything and would not divulge specifics over the phone. He insisted on meeting in person.

"Bob Barker," said Brindle-Khym, using the code name he and Qadeer gave him, "spoke cryptically about how what he had was going to be a bombshell." Barker refused to disclose where he lived, but agreed to meet at the bar of a Sheraton hotel in Philadelphia.

When Barker arrived, he seemed anxious. Throughout their conversation, Barker was vague, but he hinted at titillating evidence: "crazy parties" and "scandals." Then he demanded to be compensated.

This presented a risk. Brindle-Khym and Qadeer debated the value of the unknowable and its admissibility in court.

"We asked ourselves, 'How good is the information? How do we use it?' What are we going to say—that someone on the inside sold this to us? We had to have a way to corroborate any leaks."

Barker did a poor job of selling his secrets. He confessed he was desperate for money for his family, and his credibility was questionable. He was unemployed, had been fired from DHB, and had personal run-ins with Brooks and was frightened of him. Brindle-Khym knew that Brooks terrorized his employees.

"We knew that Brooks was unhinged, that he could pop off, was volatile," he said. "When negotiating with union officials, Brooks would descend into a stream of invective. He would yell at us through the chain-link fence at the Point Blank factory where we were picketing." Boca Raton Police Department records reflect several complaints in which people (not his employees) alleged that Brooks behaved belligerently dating back to 1995: confrontations at restaurants, beach clubs, and parking lots.

Brindle-Khym and Qadeer refused to pay Barker and considered him a false lead, but later, upon reflection of how they handled him, Brindle-Khym speculated that maybe they could have gotten more out of Barker.

"We could have sussed out what he knew and how he knew it without paying him. There are ways to corroborate information. Maybe he could have led us to other sources, guided us to what we needed."

Back at the office, Brindle-Khym was undeterred. He knew DHB was scheduled to hold its annual shareholders' meeting soon and one of the agenda items was to vote on whether to reelect the firm of Grant Thornton as its auditor. Brindle-Khym suggested that the union publicly oppose the rehiring of Grant Thornton, because the accounting firm had failed to spot the TAP transactions.

Brindle-Khym contacted Institutional Shareholder Services, a proxy advisory service that recommends how shareholders of public companies should vote on issues. ISS is influential, because it provides advice on myriad arcane topics that institutional shareholders—such as pension funds, mutual funds, and hedge funds who control large blocks of stock—don't spend time analyzing.

"Large investors can't follow every single company whose stock they own, so they pay ISS to advise them how to vote," Brindle-Khym explained.

He shared his evidence with ISS and lobbied the company to recommend that DHB shareholders oppose the reelection of Grant Thornton. ISS was persuaded. "We reframed the TAP issue as an auditing failure," Brindle-Khym said.

It worked. After the shareholders' meeting, DHB filed a Form 8-K with the SEC, which companies are obligated to do within four days of "major events," announcing that Grant Thornton had resigned.

Yet DHB disclosed no conflicts and cast the resignation as benign. But in response Grant Thornton filed its own 8-K to correct the record.

Reading from the official record, Brindle-Khym told me, "Grant Thornton said it resigned because 'it identified certain deficiencies involving internal control it considered to be significant deficiencies that, in the aggregate, constituted material weaknesses. . . . These deficiencies included the failure to disclose certain related party transactions.'"

Brindle-Khym called this "total corporate-speak," adding, "In other words, 'Everything got fucked up. Maybe there's a massive fraud. Maybe Brooks didn't tell us about TAP. Maybe our audit sucked, and we missed the fact that TAP existed. Maybe a few union gadflies and amateur financial gumshoes spotted what we couldn't see.'"

Brindle-Khym waited for Grant Thornton to call UNITE to ask for corroboration of the union's findings, but the call never came. "I was dying for them to ask me for our records," he remembers. "I would have turned over all our documentation."

Brindle-Khym says that UNITE's third complaint about the TAP fraud and Grant Thornton's resignation were the tipping points.

Brooks soon reached out to the union seeking to settle the labor dispute. UNITE agreed to stop picketing, and by the spring of 2004 a deal was reached. DHB would recognize the union at one of its plants and agreed to a modest wage increase and some improvements in health-care benefits.

The union spun the agreement to the media as a major victory. And to a certain extent, it was. "The campaign was about broader social justice," Cooper recalled. "We got to take a big swing at corporate America and corruption. And workers got a contract and some basic respect."

But the agreement had its flaws. The actual wage increase was small, Brindle-Khym recalled. "And we only got part of the company" in that the union settled on organizing one factory, while DHB kept two of its other factories operating nonunion.

"UNITE's leaders also got distracted trying to orchestrate a merger with another union, which turned out to be a failure from the UNITE perspective," said Brindle-Khym.

Brindle-Khym feels UNITE should have leveraged its new power. "We could have used the DHB deal to organize the whole industry. There's not much garment production left in the U.S. except military apparel, and this falls under that umbrella. . . . Military apparel has to be made in the United States by statute. We should have used our momentum to go after other body armor manufacturers. We settled for pennies when we could've got dollars."

As Brindle-Khym sees it, the politics of organized labor stymied its own goals. "At the higher levels of the union, they wanted to notch this as a victory and move on. They can say, 'We settled this case, we organized the workers, we got higher wages.' This garners more labor support and more money."

About three years later, in October 2007, federal criminal prosecutors in Brooklyn charged Brooks with insider trading, fraud, and tax evasion. Along with Sandra Hatfield, DHB's COO, Brooks was accused of inflating the value of the Interceptor vest to meet financial projections and failing to report to the IRS millions of dollars in bonus payments. The indictment acknowledged that the union's complaint to the SEC set the wheels in motion for what ultimately became a criminal investigation.

During an eight-month trial in Central Islip, Long Island, in 2010, labeled "uproariously and persistently unseemly" by *The Atlantic*, prosecutors laid out proof that Brooks pillaged DHB of $190 million, and pocketed $69 million in a single day.

Prosecutors and defense attorneys argued at length about Brooks's mental and emotional health. Brooks's lawyers commissioned an eminent forensic psychiatrist to examine Brooks. One of Brooks's lawyers ended up filing the psychiatrist's three-hundred-page report as part of an unrelated civil lawsuit in another court, where Brindle-Khym discovered it years later.

In September 2010, a federal jury convicted Brooks on fourteen counts of conspiracy, mail and wire fraud, securities fraud, obstruction of justice, and lying to auditors. Brooks had already pleaded guilty to conspiracy to defraud the IRS and filing false tax returns.

Brooks, according to Loretta Lynch, then the U.S. attorney for the Eastern District of New York who became the U.S. attorney general during the Obama administration, used the company as "a vehicle for plunder and a means to feed his own greed." He "fancied himself a master of the sport of kings," she said. "In reality, he was a selfish man who looted his company, defrauded his investors, lied to the SEC and the investing public, and sought to profit through insider trading right before the collapse of his house of cards."

Brooks used shareholder money to fund a "brothel tent" at a company party (this, Brindle-Khym suspected, is what Bob Barker was referring to at their meeting at the Philadelphia Sheraton), spent $35,000 on pens, and flew in Aerosmith, 50 Cent, and Tom Petty to perform at his daughter's bat mitzvah. The story was fodder for the tabloids.

In August 2013, Brooks was sentenced before the U.S. district judge Joanna Seybert. Sitting next to Brooks at the defendant's table was Gerald Shargel, whom *The New Yorker* called "one of the most brilliant criminal defense attorneys in America." (Shargel was just one of dozens of lawyers retained by Brooks during his criminal proceedings.)

Brooks made a brief statement in which he apologized to the judge and his family, but not his victims, of which there were thousands. Critics of the company, led in no small measure by UNITE, which never received an acknowledgment from the SEC that it had received the union's three complaints, exposed a blatant pattern of criminal activity. Battered by shareholder lawsuits, criminal investigations, and regulatory probes, DHB filed for bankruptcy protection in 2010.

Judge Seybert sentenced Brooks to seventeen years in prison and ordered him to pay more than $70 million in forfeits and fines.

At the sentencing, Brindle-Khym sat near the back of the courtroom taking notes. After it ended, he walked out into the hallway and waited for Brooks's legal team. Having tracked Brooks's labyrinthine trip through the criminal justice system for a decade, he was curious how his lawyers would react to the sentence.

"They were overjoyed," Brindle-Khym said: the government had asked for a thirty-year sentence. Many felt Brooks's punishment did not match his crimes, but Brooks died in October 2016 in prison in Danbury, Connecticut. He was sixty-one.

Brindle-Khym continued to follow Brooks's saga even after the trial. "The whole ordeal was bittersweet," he remembers.

"It should have been a major triumph. Here you had hundreds of immigrant workers who had no political or economic power banding together and beating a wealthy, powerful company. It could have been a place where workers lived with more dignity, had more decency in their lives, established a foothold in part of the American dream."

Brindle-Khym also laments a lost opportunity for the labor movement. "The South is not a stronghold of organized labor, but UNITE had spent decades building a presence there. All of the garment factories that worked for the military were in the South—Alabama, Mississippi, Florida."

UNITE's campaign was, in one regard, disastrous. "Instead of establishing decent wages and working conditions with devoted workers who aspired to make it in America, the company went bankrupt and all the workers we fought so hard to organize lost their jobs," said Brindle-Khym. "All of them."

It was a gory wreck. The union too has suffered, not as a result of the Brooks case, but because of political infighting; UNITE no longer exists as a stand-alone garment workers' collective.

What's more, the criminal case against Brooks unraveled in some respects. Because he died while his appeal of the jury's verdict was pending, a federal appeals court vacated most of his convictions in August 2017.

"Even in death, Brooks was one step ahead of the law," Qadeer said. Only the tax convictions, to which he had pleaded guilty and were not challenged in his appeal, outlived Brooks.

Brindle-Khym is still a staunch advocate of workers in their adversarial posture with management, but the Brooks case has left him conflicted.

"All of that struggle, all of that strife, the picket lines, the firings, indictments, the lawsuits," he said. "Sometimes I think all of it was pretty much for naught."

*B*rindle-Khym's work and his approach to the Brooks case, however, were profound. He could not have known that his work, which came at the request of DHB's own embattled employees, would ultimately imperil them.

In the past two generations, the labor movement, especially in the private sector, has lost significant power. While roughly one-third of government employees now carry union cards, only about 7 percent of their private sector comrades pay dues.

This decline in labor's membership has tracked the widening chasm of income inequality. Wealth is increasingly concentrated in the hands of fewer Americans, and corporate tax rates continue to be slashed. What's more, the social safety nets we built in the 1960s are being dismantled. Companies are winning; workers are losing.

Brindle-Khym and his colleagues sought to create a balance of power to counteract vast corporate resources. Their creativity, passion, and partisanship symbolize how the less powerful members of society might mobilize in their own defense in the absence of support from the government.

TYLER MARONEY • 244

What's more, Brindle-Khym's investigation filled a vacuum left by financial watchdogs and helped bring forward the kind of information that consumers need if they are to make informed decisions when investing in the public markets.

Companies whose securities are traded on American stock exchanges, like the Nasdaq and the New York Stock Exchange, are regulated by the Securities and Exchange Commission, which was formed in the early 1930s to prevent another Great Depression. (When it created the SEC, Congress also devised a system in which investors could sue companies for fraud and other violations of securities laws.)

Among the requirements for listed companies is that they regularly disclose their financial statements and report what lawyers call "material," or serious, information in their filings. Materiality is a broad concept and encompasses anything that a reasonable investor would consider important to know in deciding whether to invest in a company—for example, the resignation of a CFO, a major acquisition, the receipt of a criminal subpoena, or a substantive civil lawsuit.

Such disclosure is crucial because, the reasoning goes, the public must have access to issuers' independently audited, management-verified records so investors don't dump their money into frauds, mismanaged companies, or anything in between.

What Brindle-Khym knew throughout the Brooks case is that most companies work hard to disclose as little as possible. They behave as people do: they are reluctant to confess to vulnerability. And when they do open up, their descriptions of these material events—and, more broadly, of their financials—are often euphemistic, vague, or short on detail.

Another regulatory requirement is that companies alert shareholders to related-party relationships and transactions above a certain dollar threshold. Related parties are people and companies—such as corporate entities, trusts, and members of officers' immediate families—who do business with the issuer or its affiliates.

For example, if Public Company A hires Private Company B to build silicon chips that Public Company A uses to construct mobile phones and Private Company B is owned by Public Company A's founder's sister, that probably needs to be disclosed to investors and the public. Or if a company reimburses its chairwoman for flights that the CEO takes on the chairwoman's Learjet to attend marketing meetings, that relationship should probably be disclosed too.

The idea is that transparency will act as a check to deter insiders from abusing their access to the corporate treasury at the expense of the company's shareholders, who, after all, own the company.

According to a not-for-profit organization called the Financial Accounting Standards Board, which sets standards for general accounting principles, there are specific types of transactions considered "indicative" of related-party transactions, such as borrowing money interest-free, selling real estate for a price that differs significantly from its appraised value, and lending money with no terms for when the debt will be repaid.

And every public company must hire an outside auditor to

identify transactions that pose a conflict of interest. It is not the transaction that is inherently improper but the failure to disclose it. It is not illegal to hire your son's advertising firm to produce television commercials to advertise the electric cars you manufacture; it is illegal to hide that relationship from your shareholders.

Brindle-Khym, who is now my business partner at QRI, and his colleagues were well aware of how such regulations work and, by extension, how an obscure concept like an undisclosed related-party transaction could be so disastrous for DHB. Without this kind of investigative mind-set, there would be, I suspect, more CEOs like David Brooks running public companies.

EPILOGUE

One afternoon in 2016, I stood near a dock on the New Jersey waterfront preparing to raid a warehouse full of hardware manufactured by Brize Electronics, which makes televisions and mobile phones.

Six of us had trundled out of a black Suburban: two rookie cops in uniform; John Wideman, a lawyer from Kohler & Stanner in New York City, who was responsible for the stewardship of this operation; Nikky Webster, John's summer associate; Hank Danner, a former FBI agent; and me.

I wanted to be first to crash through the steel doors, but because I was carrying the iPhone (not the Glock or the seizure order from the court or the locksmith's tools), I fell back.

The leads that got us here were some intelligence from someone who previously worked at Brize and whom we had developed as an informant, the results of some controlled buys at a retailer in Newark where we suspected counterfeit phones were sold to consumers, and two weeks of surveillance at this warehouse.

It was enough to persuade a federal judge to authorize us to search, seize, copy, and sequester from this warehouse any products labeled "Brize," business records, computers, and any cash over $10,000.

The squat brick building was in a desolate, industrial neighborhood near Bayonne. We told our driver to wait conspicuously in front of the building and keep the motor running.

I braced for what I hoped would be something spectacular. Infrared beams that slice through the darkness. Chaotic shouting and calls for surrender echoing through the cavernous warehouse. We'd gang-tackle suspects and interrogate them while they wept in folding chairs under the dim bath of a naked lightbulb that hung from the rafters.

This raid was one of many we undertook for Brize, which understood that finding fraudsters and seizing counterfeit goods would not halt fakes from flooding the U.S. market. It was, we were told, a "whack-a-mole" approach designed less to solve a problem than to send a message to the bad guys: we won't stop coming for you.

Private investigators are hired at different times by different clients for different reasons, appearing in unanticipated points throughout a project's evolution.

Sometimes we are present at creation—and help anticipate threats (by warning that employees may steal from their employers using a corporate American Express card) and plan strategy (by setting spending limits or monitoring email traffic).

Sometimes we amble in, unaware of what preceded us or what is to come—and, at the very least, provide some measure of insurance (by finding an offshore bank account that will be garnished to repay an old debt).

Sometimes we are a Hail Mary—and either avert disaster or emerge triumphant (by persuading a witness who lied under oath for the prosecution in a criminal trial to recant her statements).

It is not uncommon to be hired more than once by the same client for the same reason—for instance, to investigate vendor fraud at a company with operations in many countries. Not all problems can be solved once.

But clients would rather not have us grouped among Favorites in their iPhone contacts. I've been told more than once by appreciative clients that they hope they never see my face again. After all, our phones often ring not when smoke is slipping under the door but when flames are leaping out the windows.

But the firefighter metaphor is not quite apt. In fact, among my goals in this book has been to demonstrate that our services are necessary throughout the ecosystem of global transactions and disputes and that we must therefore use legal tactics on behalf of our clients.

We prepare trial attorneys to file complaints and depose expert witnesses, and arm embattled public companies to defend against activist shareholders. We help criminal defense lawyers impugn the credibility of government witnesses, help corporations prevent the theft of trade secrets, and assess electronic security threats. We unmask hackers, prepare investors for acrimonious negotiations, track and seize the assets of irresponsible debtors, and expose weaknesses in global supply chains.

The intelligence that Julian Fisher gathered in Kenya suggested that terrorist attacks in Nairobi were not targeted but random. This information gave comfort to a nervous client who was considering divesting from swaths of an entire continent.

In exposing the corruption of an elected official in Connecticut, Jim Mintz brought closure to a scandal that was national news, and he brought a measure of transparency to a broken and murky government contracting system.

Luke Brindle-Khym unveiled the misdeeds of a wealthy executive in the service of providing a check on corporate greed and supporting the powerless. His work filled gaps left by federal law enforcement officers and regulators.

My discovery that David Dasha had a hidden past might have prevented an investor from losing millions of dollars.

Ian Casewell kept a media giant honest in uncovering real "fake news."

The work we do is focused, specific, and ultimately revelatory about the choices people make.

As we stood outside the warehouse on the New Jersey waterfront, we readied our approach.

Hank Danner, the former FBI agent—who is bald, red-faced, and barrel-chested and whose last gig was infiltrating Islamic charities in Brooklyn—was exhilarated.

"There's no thrill like hunting another human being," he said to me. His American flag cuff links caught the morning sunlight.

"What?" I asked, not sure I'd heard him correctly.

"Ernest Hemingway."

"Ernest Hemingway?"

"He's a writer."

I have read Hemingway; I just didn't expect to hear him quoted during an operation on the New Jersey waterfront.

The stevedores inside could not have known that we were coming. The first police officer rang the doorbell. The second unclipped his holster. Wideman pulled the court order from his breast pocket. Webster readied a notepad and pen. Danner unsheathed his picks, tweezers, and turning tools. And I pressed the video-record button on my camera.

The raid was less than sensational. A warehouse manager casually opened the door and politely invited us in. We collected the merchandise and documents while chatting with the bored employees. (There was no cash and no computers.)

I recorded the operation, uploaded the files to our server, then spent the afternoon noting serial numbers on products that would help our client determine if the goods were counterfeit.

What I had imagined would be a dramatic search, seizure, and arrest—like something out of a dramatic film—turned out to be a rational, reasonable, by-the-book operation.

For all the mission's tediousness, it proved worthwhile. Using the evidence we collected at the warehouse, Brize determined that 30 percent of the products we seized were fake. We had discovered the aorta of this counterfeit operation in the United States. Our next assignment would be to track the fakes back to a plant in Vietnam.

There is an absurd and wondrous quality to this job. After all, corporate private detectives interact, sometimes during a single project, with odd characters—corporate titans, lawyers learned and lawless, racketeers, grifters.

In the final analysis, it is our versatility, perhaps, that keeps us employed: We are painstaking and nimble, law-abiding and enterprising, imaginative and deliberate—all to ensure our clients get the hidden information they need. We are lubricant, bandage, and weapon.

ACKNOWLEDGMENTS

Writing this book was like working a complex investigation—it forced me to take many unanticipated turns and allowed me to consult with many remarkable people.

Jules Kroll, Jeremy Kroll, and Dana Kroll offered me my first job as a private detective, trained me, and unwittingly altered the course of my life.

Jim Mintz, Marc Fader, and Chris Weil also took a chance on me and heaved me forward.

Luke Brindle-Khym taught me that obsession is a virtue, sometimes you have to take sides, and this gumshoe game should be fun.

Annie Cheney inspires me with her creativity and work ethic—as does the entire staff of QRI.

Dozens of people contributed to this book anonymously. Here is my gratitude.

When I was a journalist, I became acquainted with three reporters, towering figures all, whose lifework and brief encouragement were more profound than they know: Jon Lee Anderson, Martin Smith, and Robert Krulwich.

The following, in alphabetical order, deserve special thanks for their contributions or support: Taline Al Assad, Jessica Bendinger, Rishi Bhandari, Will Bourne, Robert Capper, Ian Casewell, Suzanne Clarke, Isabelle D'Ursel, Wendy Davis, Christopher Faherty, Hector Feliciano, Julian Fisher, Kevin Flanagan, Shaun Gatter, Charles G. Geyh, Julian Grijns, Daniel Hall, Maria Harris, Matthew Kaplan, John Knapp, Justin Lane, Emma Lindsay, Charlie Linehan, Denman Maroney, Hagen Maroney, Erin Martin, David Matthews, Chris McCavitt, Eleanor Newirth, Joseph Newirth, Michael Newirth, Marcus Oliver, Sarah Oliver, Laura Perciasepe, Ahmer Qadeer, Bob Randall, Romesh Ratnesar, Michael Redman, Stefan Reich, James Rigby, Brian Rosenthal, Lauren Sandler, Melchior Scholler, Nitul Shah, Polly Sprenger, Valerie Steiker, Nicholas Stein, Nik Steinberg, Albert Togut, Kevin Treuberg, and Chris Young.

I am thankful for Ellis Levine's sage legal advice.

My college history professor Gary Ostrower taught me to be skeptical and write with clarity. I hope he approves of these pages.

Jenny Li is a talented lawyer and researcher whose additions to the manuscript were invaluable.

Daniel Smith and Jonathan Tepperman improved my turns of phrase, word choice, reasoning, and the structure of many chapters.

Nell Casey, Anne Diebel, and Gemma Sieff contributed thorough, crucial edits and refused to hold back their opinions.

Jake Morrissey, my editor, has a pleasant and compassionate demeanor but an unrelenting red pen. Thank you, Jake, for both.

Geoff Kloske, my publisher, once warned me he would not line edit my manuscript. Instead, he would "hum a few bars" for me to imitate. If any melodies or harmonies can be heard in this prose or these ideas, they are as much his as mine.

Karen Newirth, you show me every day that dreams are meant not simply to be achieved but to be layered on top of each other, improved upon, and wrapped around each other. This book was a dream, like so much else in my life, until you arrived.

INDEX